Sacred Groves, Cultural Ecosystems and Conservation

Sacred Groves, Cultural Ecosystems and Conservation

Edited by

Rena Laisram

Cambridge
Scholars
Publishing

Sacred Groves, Cultural Ecosystems and Conservation

Edited by Rena Laisram

This book first published 2023

Cambridge Scholars Publishing

Lady Stephenson Library, Newcastle upon Tyne, NE6 2PA, UK

British Library Cataloguing in Publication Data
A catalogue record for this book is available from the British Library

ISBN (10): 1-5275-0106-X
ISBN (13): 978-1-5275-0106-5

To

Kris Bani

&

Jeevan

CONTENTS

ACKNOWLEDGEMENTS

First and foremost, I would like to acknowledge Cambridge Scholars Publishing, UK for giving me the opportunity to be part of this exciting project. My pleasant experience with CSP at the publication of my book, *Religion in Early Assam: An Archaeological History* (2019) encouraged me to embark on this journey once again. I express my appreciation to the friendly and cooperative staff of CSP for their constant support and help.

This volume would not have been possible without the scholarly contribution of the authors who evinced interest in the theme of the book. I would like to thank each and every one of them for their time and expertise. The overwhelming response has been a motivation to take up the responsibility as a Guest Editor.

My colleagues and students at the Department of History, Gauhati University, Assam, India have contributed to stimulating academic interactions which has encouraged me to explore new fields of research. I am thankful to them.

I remember my loving parents *Ima* (Mom) and *Pabung* (Dad), Rajkumari Kadambini Devi and Laisram Gopal Singh, who have been the best examples of dedication, hard work, and sincerity.

My husband Jeevan, and sons Bani and Kris have always been my greatest strength.

The unprecedented pandemic times which created a global health crisis like never before has had a deep impact on the way we conceive the world. The theme of this book finds immense relevance in the context of biodiversity conservation and Sustainable Development Goals.

This work is a collaborative effort of so many people who have faced the challenges of uncertain times and emerged stronger in their own ways. My deepest gratitude to all of them.

Rena Laisram

LIST OF ILLUSTRATIONS

Map

Figures

ABBREVIATIONS

FSI: Forest Survey of India
GFA: Gompa Forest Areas
IJTK: Indian Journal of Traditional Knowledge
ISFR: India State of Forest Report
ICCA: Indigenous and Community Conserved Areas
ICOMOS: International Council on Monuments and Sites
IUCN: International Union for Conservation of Nature
NCMM: National Commission for Museums and Monuments
NEC: North Eastern Council
NTFP: Non-Timber Forest Products
SFR: State of Forest Report
SGS: Sacred Groves
TFAP: Tanzania Forestry Action Plan–North Pare
TNA: Tanzania National Archives
TKS: Traditional Knowledge Systems
TNT: The Northeast Today
UNESCO: United Nations Educational, Scientific and Cultural Organization
UNDRIP: United Nations Declaration on the Rights of Indigenous Peoples

Introduction

Rena Laisram

Concept and Context

Ecosystem degradation represents one of the major global concerns which is threatening human well-being and biodiversity as a whole. Although nature worship has been integral to religious belief systems around the world, forest covers have been shrinking at an alarming rate due to disregard for indigenous cultures and encroachments in the name of development. *Sacred Natural Sites* are part of a broader set of cultural values that different social groups, traditions, beliefs or value systems attach to places and which 'fulfil humankind's need to understand, and connect in meaningful ways, to the environment of its origin and to nature' (Putney 2005, 132). The close relationship between the divine world and the natural environment is traced to Graeco-Roman antiquity which is attested by textual and material evidence (Caroll 2017, 3). The terms *alsos* and *temenos* were used in Greek literature to refer to a terrain set aside for a divinity (Barnett 2007). Natural sacred sites are found in all the continents except Antarctica (Bhagwat and Rutte 2006), Africa and Asia being home to a large number of sacred groves. The size of sacred groves may range from a few trees to dense forests of hundreds of hectares. Here, it is crucial to distinguish between a single tree being regarded as sacred, such as the famous oak of Zeus at Dodona, in north-western Greece (Frazer 1925), and a sacred grove, which refers to a cluster of trees that are close together and accorded sacrality. The type of evidence available includes texts, visual representations, and material remains.

Hughes and Chandran (1998) define sacred groves as...

> ".... segments of landscape containing vegetation, life forms and geographical features, delimited and protected by human societies under the belief that to keep them in a relatively undisturbed state is expression of an important relationship of humans with the divine or with nature."

Sacred groves are relatively undisturbed tracks of virgin forests with often very old trees and are categorised according to the deity, shrine, burial or cremation grounds associated with it. In tribal-based societies, sacred groves provide spaces for veneration of ancestors through a chain of rituals. However, irrespective of particular belief systems, sacred sites are recognised as community-based symbols of spiritual and socio-economic values. Myths and taboos have attempted to explain the relation between the spiritual and the human world. A storehouse of biodiversity, such sacred spaces are protected by the communities since their ethnic history is closely connected to it through mythological traditions. Women and men experience sacred space differently and often play contrasting roles within the spaces. It must be noted that most of the world's religions have promoted a patriarchal structure of male authority and female subordination, and the liturgical practices and rituals are performed by male functionaries. However, the Okinawas in Japan offer an example of an indigenous religion where women lead the official mainstream religion of the society (Sered, 1999). Religion's central role in consolidating gender difference and inequality was recognized, explored and critiqued by 19th century feminists such as Elizabeth Cady Stanton (*The Woman's Bible*, 1999). Given the enormous influence of religion in human affairs, the continued exclusion of women's roles of religion is bound to reflect negatively on Sustainable Development Goals.

The *United Nations Declaration on the Rights of Indigenous Peoples* (UNDRIP, 2007) addresses the protection and recognition of sacred natural sites. There are currently thirty-six biodiversity hotspots worldwide with a unique set of diverse ecosystems that are home to more than 2 billion people. Conservation experts and ecologists have taken increasing interests in these sites, and acknowledged 'the ability of traditional societies to preserve biodiversity through their knowledge and rituals' (L'Homme 1998, 10). The United Nations declared 2010 as *The Year of Biodiversity*, bringing to focus the value of biodiversity and urgent need for action. The two key organisations that play a crucial role in addressing indigenous concerns of such invaluable sites are: *International Union for the Conservation of Nature* (IUCN) working in the field of nature conservation and sustainable use of natural resources, and *International Council on Monuments and Sites* (ICOMOS) which focuses on the conservation and protection of cultural heritage places. The need to protect sacred groves is agreed upon, but the question is: what urgent steps are being taken by individuals, governmental and non-governmental organisations to safeguard the endangered environment?

Conserving Sacred Groves: An Interdisciplinary Approach

This volume addresses cultural and conservation aspects of sacred groves threatened by habitat loss through an in-depth study of representative samples from Africa and Asia. Based on the long-term research of the contributing authors, the nine chapters reflect a continuous process of redefining the sacred grove within an interdisciplinary framework grounded on existing literature and ethnographic field research. In analysing the various dimensions of the sacred sites, the researchers have also collected information through interview and questionnaire methods. The addition of photographs, tables, and figures have enriched the subject under discussion. Biodiversity hotspots around the globe are homes to culture-specific vegetation, which offers a huge scope for scientific research. The highlight of the discourse is the complex interactions and negotiations between the 'sacred' and the 'secular' - which brings center-stage the subject of sacred status that communities have given to nature.

The book is organised in two parts, viz.,
Part I: Culture, Nature and the Sacred;
Part II: Sacred Ecology and Sustainable Development.

Chapter I: *Sacred Groves of the Karbis in Religio-Cultural Perspective* by Laxmi Hansepi, examines the centrality of sacred groves in Karbi society and attempts to gain insights into the changing dynamics of interface between nature and culture. It is worthwhile to mention that culture is understood as that complex whole which includes knowledge, belief, art, morals, law, custom, and any other capabilities and habits acquired by man as a member of society. The sacred groves of Karbi Anglong district (undivided), located in Assam, India exemplify the role taken by communities in preserving natural resources that are consistent with Indigenous and Community Conserved Areas (ICCAs) globally. The Karbis as one of the largest ethnic groups in Assam represent one of the many Indo-Mongoloid groups of people whose worldview centres around ancestor worship. The community shares a close connection with the ecosystem culturally and/or because of livelihood needs. Animistic belief system is manifest in the living tradition which is reinforced through various rituals performed throughout the year. All members of the clans partake the feasts and revive their bonds of kinship. The myths and taboos ensure that the sacred groves are protected and maintained. The study reveals that the community is the major decision maker in management and conservation of the natural resources. Therefore, the community-conserved areas thrive on a symbiotic relationship between nature and culture.

In ***Chapter II:*** *Ritual Practices and Nature Conservation among the Misings of Assam,* Ripun Doley discusses the Mising religious belief system in the context of ritual observances and conservation of nature. Ritual as a component of culture is crucial to the sustenance of social organisations among ethnic groups. The importance of rituals in religious activities have been explored by anthropologists such as Victor Turner, who defined 'ritual' to mean 'prescribed formal behaviour for occasions not given over to technological routine, having references to belief in mythical beings or powers.' Prescribed ways of carrying on religious acts and procedures, ways of worshipping, praying, chanting, sacrificing, making offerings etc., are practiced in rituals. Ritual is an inevitable component of culture, extending from the large-scale social and political processes to the most intimate aspects of self-experience. The Misings of Assam follow social norms which can be divided into four categories, viz., beliefs, rituals, superstitions, and taboos. Taboo is a ritual prohibition that is rooted in specific cultures. It is through the practise of these customs that the community has helped to maintain the sacred sites as repositories of traditional knowledge systems. The traditional priests also play a key role not only in officiating as priests, but also as healers or medicine - men as the indigenous method of curing diseases has been known to them and handed down from generation to generation.

Chapter III: *Of Groves and the Goddess: Women, Ecology, and Sacredness* by Sakti Sekhar Dash, analyses the complex dynamics between human community, religion, socio-cultural beliefs, and ecology. It examines the role of groves and the creation of ecological sacredness which coincides with the emergence of the cults of the *magna mater*, celebrated as the Great Goddess of the Earth.

Rosy Yumnam explores textual knowledge to understand nature and culture in ***Chapter IV:*** *The Tunes from the Sacred Groves: An Ecocritical Reading of Hijan Hirao.* Ecocriticism is one of the crucial recent research trends in the field of cultural and literary studies. The main aim and concern of ecocriticism is to explore the artistic and creative power of literature and to relate it to the environment in the present times. Relatedly, cultural ecology is an approach of ecocriticism which aims to unify various cultures of knowledge across diverse disciplines and sub-disciplines. This chapter seeks to examine the ecocritical elements in *Hijan Hirao*, a folk song of Manipur in North-East India. The framework of cultural ecology is explored to read the folk song. *Hijan Hirao* is a long narrative poem which is sung on the closing day of the *Lai Haraoba* festival of Manipur. The sacred grove (*umang*) occupies a pertinent place in the religious landscape of the Meitei community of Manipur. Revered deities called the *umang lais* (sylvan deities) are believed to reside in the sacred groves. In the *Lai Haraoba* festival, the *umang lais* are given due respect and homage through various songs, dances and rites and rituals. *Hijan Hirao* is one such song performed during the festival. The sacred song is in the form of a lament of the 'tree parents' for their loving 'tree son.' As ecocritics dwell on the role of conserving ecology through texts, the chapter aims to look at how the tunes of *Hijan Hirao* can inculcate the noble thought of conserving the environment and culture in the minds of the readers. Human culture is considered to be mutually dependent on natural energy cycles and ecological processes, so the implication of culture through the myth narratives of *Hijan Hirao* endears to bring about ecological awareness among the community. Ecocritics espouse the role of nature in texts concerning human culture and nature and how these texts play a critical role in conserving ecology. The tunes of *Hijan Hirao* through its ethical orientation towards nature build a formidable task of preserving and conserving the environment.

Chapter V: *Biodiversity Conservation and Ecosystem Services Provision of Sacred Groves in Africa* by Jonathan C. Onyekwelu, examines the biodiversity, religious and cultural, socio-economic, ecosystem services provision of sacred groves in the context of Africa. It also discusses the potentials of sacred groves in sustaining the provision of their important values in the face of current anthropogenic threats. Sacred grove system evolved through the interactions between indigenous communities and their environments; they act as connecting link between man, nature, horticulturalist heritage, religious and ethnic beliefs. Their success in delivering ecosystems goods and services to communities is hinged on their multifaceted dimensions and functions. Today, sacred groves have become integral part of rural lives and are revered for their religious, cultural, medicinal, social, economic, psychological and educational functions. However, some of these important fulcrums of sacred groves are not only currently threatened, there are evidences that they are being eroded. Given the threats, the extent to which sacred groves would continue to provide their important functions now and in the future is uncertain. Some strategies towards mitigating the threats and repositioning sacred groves have been suggested.

Chapter VI: *Capital Conversion Institutions: Sacred Groves in Northern Tanzania* by Michael J. Sheridan, is a Tanzanian case study in which the author argues that ecological, social, political, and cultural dynamics intersect and disconnect over time in African sacred groves. Because "sacredness" is a contested process among unequal social actors, conservation policy for sacred groves should anticipate complexity and focus on adaptive co-management at local, national, and global levels.

Chapter VII: *Cultural and Ecological Services from Urban Sacred Green Spaces in Bhubaneswar, India* by Antaryami Pradhan, Alison A. Ormsby, and Madhusmita Samal, analyses the perceptions of visitors regarding provisioning of cultural ecosystem services in the urban sacred green spaces in Bhubaneswar, India. Urban green spaces have particular values because they offer the benefits of ecological and cultural services to city dwellers and visitors. The green spaces in cities are at risk due to rapid urbanization and conversion into built areas. Past researches have documented the importance of green spaces in urban areas. However, limited studies exist that measure the value of urban sacred green spaces in provisioning of cultural ecosystem services. An interview was conducted with 325 individuals visiting the sacred green spaces of Bhubaneswar and they were asked about their perceived benefits from and purposes for use of these spaces. Approximately one third of the respondents identified religious purpose as their cause of visit because of the cultural significance of the site. The second largest purpose for visiting the sites was for health benefits and physical activities, such as yoga, walking, or exercise, and mentioned by 31.5 percent of respondents. The main perceived benefit from the sites was recreation, identified by 45 percent of the interviewees. Despite identified benefits, there is an increasing concern among visitors regarding the planning and management of these sites for better provisioning of their ecological and cultural services. The result of this study has the potential to influence management to improve the benefits received by visitors of the sites and to inform city planners for establishment of future green areas, particularly in sacred spaces, understanding that there are many reasons that people visit these sites.

Parismita Borah examines the sacred groves of Meghalaya as a repository of flora and fauna in ***Chapter VIII:*** *Safeguarding Sacred Groves in Meghalaya: Issues and Challenges.* It attempts to analyse the issues and challenges faced by sacred groves to safeguard the threatened ecosystem. It will also gain insights into the role played by the government for biodiversity conservation and management of the forests in the state. The priests or *lyngdohs* have been given the responsibility by the district councils of the state to nurture and preserve the sacred natural sites. This is an encouraging step as the partnerships of the government and local specialists is a positive way to share knowledge and expertise for the common cause of safeguarding the ecosystem. An important dimension of the discussion is the role of women in the sustenance of the rich ecosystem which is quite uncommon in other parts of India or across the globe. This is on account of the matrilineal system and implications of gender role prevalent among the major ethnic groups in Meghalaya. It may be mentioned that sacred groves in general prohibit women from entering for rituals and other purposes as they are considered impure. Women and the sacred groves of Meghalaya is a topic that must be explored further in future researches.

Chapter IX: *From Sacred to Secular: Sacred Groves of Northeast India in Transition* by Meeta Deka, deals with the concept of 'sacred groves' which varies the world over as it is borne out of diverse historical, ecological, economic, socio-cultural and even political factors that shape a particular culture. This essay traces how the so-called sacred groves of Northeast India, known by diverse indigenous names, acquire newer dimension through various phases of history and hence are not static in time, space and meaning. The connotation of the term has been in a constant state of flux. This essay argues against a spiritual or religious foundation as the basis of what is 'sacred' but which evolves into a sacred space subsequently. It was undermined under British cultural hegemony and soon became a site for economic exploitation, while post-Independent Northeast India saw a spurt of ethnic movements in the late twentieth century when sacred groves became an integral part of 'shared pasts' in asserting ethnic identity. In the twenty-first century and after, with growing concern for global warming, sacred groves of Northeast India assumed the status of a symbol of conservation of biodiversity and ecology and a tourist destination!

The volume titled, *Sacred Groves, Cultural Ecosystems and Conservation* addresses a subject of increasing contemporary relevance as ecosystems are being depleted at an alarming rate worldwide. The purpose of this collection of essays is to bring together different perspectives on sacred groves in the context of the cultural and spiritual dimensions of biodiversity conservation. In offering an experience of sacred natural sites in varied cultural contexts of Africa and Asia, it raises a common concern for natural resource management. It may be noted that a scientific understanding of harmony and co-existence of religion and environment will go a long way towards a sustainable ecological balance. As Rappaport (1999) aptly remarks: 'Humanity' s task today is to assume responsibility for life as a whole on this planet and religion, the synthesis of objective law and subjective meaning, is indispensable to that end.'

References

Barnett, R. 2007. "Sacred Groves: Sacrifice and Order in the Landscapes of Ancient Greece." *Landscape Journal* 26(2): 252-269. Madison, Wisconson: University of Wisconsin Press.

Bhagwatt, Shonil A. and Claudia Rutte. 2006. "Sacred Groves: Potential for Biodiversity Management." *Frontiers in Ecology and the Environment* 4(10) (December): 591-524. Accessed November 20, 2022. 10.1890/1540-9295(2006)4[519: SGPFBM]2.0.CO;2.

Biodiversity Hotspots: The Red List Project. Accessed December 10, 2022. https://www.theredlistproject.org/hotspots.

Carroll, Maureen. 2017. "The Sacred Places of the Immortal Ones: Ancient Greek and Roman Sacred Groves." In: J. Woudstra, J. and C. Roth. Eds. *A History of Groves*. London: Routledge. Accessed December 20, 2022. https://doi.org/10.4324/9781315561066-2.

Chandran, Subash M.D. and J. Donald Hughes. 2000. "Sacred Groves and Conservation: The Comparative History of Traditional Reserves in the Mediterranean Area and in South India." *Environment and History* 6(2) (May): 169–186. Accessed December 13, 2022. http://www.environmentandsociety.org/node/3043.

Frazer, Sir J.G. 1925. *The Golden Bough: A Study in Magic and Religion*. London: Macmillan and Co.

Hornborg, Anne-Christine. 2008. "Protecting Earth? Rappaport's Vision of Rituals as Environmental Practices." *Journal of Human Ecology* 23 (4): 275-283 (October). Accessed November 15, 2022. https://doi.org/10.1080/09709274.2008.11906081.

Hughes, J.D. and M.D. Subash Chandran. 1998. "Sacred Groves around the Earth: An overview." In: P.S. Ramakrishnan, K.G. Saxena, and U.M. Chandrashekara. Eds. *Conserving the Sacred for Biodiversity Management.* New Delhi: Oxford and IBH Publishing Co.

International Council on Monuments and Sites (ICOMOS). Accessed November 25, 2022. https://www.icomos.org/en/about-icomos/partner-organizations.

L'Homme, Cristina. 1998. "The Destiny of Sacred Sites." In: *Sources: Conceiving the 21st Century*. UNESCO Newsletter No. 106 (November).

Martin-Ortega, Julia, Dylan M. Young, Klaus Glenk, Andy J. Baird, Laurence Jones, Edwin C. Rowe, Chris D. Evans, Martin Dallimer, and

Mark S. Reed. 2021. "Linking Ecosystem Changes to their Social Outcomes: Lost in Translation." *Ecosystem Services* 50 (August): 101327. Accessed December 19, 2022. https://doi.org/10.1016/j.ecoser.2021.

Putney, A. 2005. "Building Cultural Support for Protected Areas through Sacred Natural Sites." In: J. McNeely. Ed. *Friends for Life*, Gland, Switzerland: International Union for Conservation of Nature.

Rappaport, R. 1999. *Ritual and Religion in the Making of Humanity.* UK: Cambridge University Press.

Sered, Susan. 1999. *Women of the Sacred Groves: Divine Priestesses of Okinawa.* New York: Oxford University Press.

Stanton, Elizabeth Cady. 1999. *The Women's Bible*. Amherst, New York: Prometheus Books.

The United Nations Declaration on Rights of Indigenous People. 2007. Resolution No. 61/295. Department of Public Information, United Nations.

Wild, Robert and Christopher McLeod. Eds. 2008. *Sacred Natural Sites: Guidelines for Protected Area Managers*. Gland, Switzerland: International Union for Conservation of Nature.

PART-I

CULTURE, NATURE AND THE SACRED

CHAPTER I

SACRED GROVES OF THE KARBIS IN RELIGIO-CULTURAL PERSPECTIVE

LAXMI HANSEPI

India is recognised as one of the thirty-six biodiversity hotspots worldwide, being host to highly endangered eco-regions that are severely threatened by habitat loss and other human activities (1). The Indo-Burma region, one of the largest hotspots, covers Myanmar, Thailand, Cambodia, Vietnam, Laos PDR and also includes the Gangetic plains, areas around the Brahmaputra River and parts of Andaman and Nicobar Islands. It is important to note that ethnic communities worldwide are mostly found to have preserved a range of natural habitats of sacred importance in the name of religion or culture (Hughes and Chandran 1998). As such, sacred groves represent a vital gene pool which promotes the conservation of many endangered species on the verge of extinction (Gadgil and Vartak 1976). They exist in all parts of the world under different names and are found in countries like Ghana, Japan, Morocco, Ethiopia etc. A subject of immense interest to anthropologists, biologists, social scientists, and policy-makers etc., sacred groves as communally protected forest fragments may range from groves with conspicuous presence, in terms of expanse, economic importance of specific species etc. to small groves (mostly < 1 ha) of local landscapes (Ray and Ramachandra 2012).

The sacred groves of Northeast India are spread across the states of Assam, Arunachal Pradesh, Meghalaya, Manipur, and Sikkim. It has a rich forest cover of more than 60% of the total area of the region, and comes under the Indo-Burma biodiversity hotspot. Largely inhabited by Indo-Mongoloid group of people, they have a close relation with nature through the religious belief systems (Kandari et al. 2014). The ethnic communities are dependent on the forest resources for their livelihood and, therefore, attach religio-cultural significance to the sacred sites (Devi et al. 2017). There are two hundred and thirty-two sacred groves in Assam with the highest number, of forty-two, recorded in Tinsukia district. The undivided Karbi Anglong district (2) has seven sacred groves (*thans*). However, only *Ronghang Rongbong*, located in Hamren sub-division in the present-day West Karbi Anglong district has been documented so far. Apart from the Karbis (3) who form the major group, the social demography of the district also includes the Garos, Tiwas, and Dimasas. The Karbis also reside in other places in Assam such as West Karbi Anglong, North Cachar Hills, Kamrup Metropolitan, Kamrup Rural, Morigaon, Nagaon, and Sonitpur.

Objectives

The sacred groves of India are examples of Indigenous and Community Conserved Areas (ICCAs) that are consistent with ICCAs globally (Berkes 2007). As with the Karbis of Assam, the community shares a close connection with the ecosystem culturally and/or because of livelihood needs. In this context, the community is the major decision maker in management and conservation and may even enforce regulations (Pathak et al. 2004). The Community Conserved Areas thrive on a symbiotic relationship between nature and culture. It is worthwhile to mention that culture may be understood as that complex whole which includes knowledge, belief, art, morals, law, custom, and any other capabilities and habits acquired by man as a member of society (Tylor 1871). This study attempts to examine the centrality of sacred groves in Karbi society and gain insights into the changing dynamics of interface between nature and culture.

Study Area

The universe of study is Karbi Anglong district (undivided) in Assam with a total area of 10,434 sq. kms, and having a topography which includes hilly terrains, and numerous rivers. It may be noted that West Karbi Anglong district was carved out of the undivided Karbi Anglong district in 2016, and today there are two districts, viz., Karbi Anglong and West Karbi Anglong. The physiography is divided between hills and plains, and as 85 % of the district is covered by hills, it is referred to as a Hill District. Some important rivers of the district include Kopili, Amreng, Borpani, Kolioni, Dhansiri, Dikharu, Longnit etc. The district is situated in the central part of Assam and shares borders with Nagaland state and Golaghat district in the East and Hojai district in the West. The South is bounded by Golaghat and Nagaon district, whereas North Cachar Hills and Nagaland state share their boundary in the North.

Sacred Natural Sites

Sacred Natural Sites (Verschuuren 2010, 2) are part of a broader set of cultural values that different social groups, traditions, beliefs or value systems attach to places. The people who hold these places to be sacred believe them to be occupied or constituted by spirits or deities which have supernatural powers, such as that of curing illness or bring rain. In Japan, for example, the sacred mountains are believed to be an embodiment of *kami* (Shinto spirits or gods). According to the *International Union for Conservation of Nature*, sacred groves, a form of nature worship, are considered as Sacred Natural Sites. (Ray n.d.). The origin of sacred groves in India were associated with hunting-gathering societies (4). However, scholars believe that with the advent of agriculture, patches of forests came to be set aside for cultivation and sacred values attributed to them (Negi 2005). Ecologically, sacred groves help preserve soil and water apart from protecting biological biodiversity. Hunting, cutting of trees and plants, and entering the site are generally prohibited, the natural ecosystem is protected. These sacred groves also help preserve age-old customs and traditions, beliefs, and practices that otherwise would have been lost in this age where globalization has threatened many indigenous beliefs and practices (Hughes and Swan 1986).

Malhotra et al. (2001) stated that in India, the tradition of Sacred Groves is reported from most parts of the country, and approximately 13,270 sacred groves have been documented so far. Sacred Groves are classified into three broad divisions (Ray and Ramachandra 2010; Roy 1998) which are as follows:

1. *Traditional Sacred Groves*, where the territorial deity is revered and are represented by an elementary symbol. Traditional groves have higher repository of reserved areas.
2. *Temple Groves* are located around the temple premise and conserved by the devotees. Temple groves have sacred plants species which could be age-old tree species historically and mythically linked with the epics, gods and goddesses. In temple groves, the leaves, fruits, and flowers are considered sacred and are offered to the deity. Interestingly, these plants are considered sacred by the local communities, who view them as incarnations of the gods and goddesses they worship. Sacrificial elements are observed widely in temple groves dedicated to a female deity.
3. *Burial Groves* are located around the burial ground and are preserved because of their beliefs.

The highest number of sacred groves in India are temple groves, followed by traditional and burial groves. However, traditional groves record higher conservation status. In India, most sacred groves are reported from regions like Western ghats, Central India, and North-Eastern India (Gadgil and Vartak 1976). Literary sources of ancient India are replete with stories and myths of the opulence of resources, viz., flora and fauna in forest areas. *Rishis* and *sages* live in the forests for meditation away from the hustle of everyday life (Malhotra 2001, Pungetti et al. 2012). Even texts such as the *Mahabharata* and the *Ramayana* mentions many rivers, mountains, and sacred forests with descriptions reflecting the importance of nature. There has always existed a symbiotic relationship between man and nature.

Karbi Religion and Culture

The traditional religious philosophy of the Karbis is based on polytheism, animism and ancestor worship. With the influence of Hinduism in the past few decades, they have come to follow the essence of both the religions (5). The spiritual and moral values of the Karbis are reflected vibrantly in their folklore and traditions. The Karbis view the natural environment as the abode of spirits. Thus, every aspect of nature is holy. The Karbis have always believed in the concept of *Arnam Keso*, that is, harmony with nature. According to the *Census of India: 2011* (6), the percentage of the Hindus in Karbi Anglong is recorded as constituting 80.10% of the total Karbi population. The centrality of sacred groves in the lives of the people reflect their deep sense of reverence for nature. It may be mentioned that the Karbis are hills dwellers and lived in proximity with the forests. They believe that every animate and inanimate object possesses a soul and, therefore, nature must be protected. In fact, the Karbi religious belief system places great emphasis on the concept of soul (*akarjong*), and believe that the soul is immortal and walks alongside humans in the form of shadows. The Karbis believe in several deities, which are both malevolent and benevolent. Collectively they are referred to as *Arnam*. The veneration of ancestors (*Tirim*) is done to appease them and seek their blessings. The ancestors are brought alive through the many community festivals throughout the year, which ensures that kinship ties are strengthened.

The Karbis worship many gods and goddesses, the holy trinity occupying the highest position in the pantheon. *Hemphu, Mukrang* and *Rasinja* are the three deities whose names are evoked in all prayers. However, amongst the three deities *Hemphu* is considered the supreme God. Apart from the Trinity, the other gods worshipped by the Karbis include *Peng, Arnam*

Kethe, Rek-anglong, and *Arnam Pharo*. *Peng* protects the household or a family from misfortune through the *peng-ase,* a ritual performed by the people at the individual or household level. *Arnam Kethe* is a house god and worshipped once in three years by sacrificing a pig. *Rek-anglong* is a household deity but the rituals are performed outside in the field, once a year, with only male members being allowed to partake the food items offered to the god. *Arnam Pharo* (Hundred God) is so named as it takes a hundred shares of rice and areca-nut. The Karbis usually decide to perform a ritual when they encounter ill-health or misfortune as they believe that bad omen occurs due to improper and deficient propitiations to Gods, and based on the circumstance, the respective gods are propitiated (Das 2011).

The Karbis practice agriculture-based economy and traditionally the land is owned by the village head known or *Sarthe* (Teron 2008, Pereira et al. 2017). As the forest provides them with food and other produce for their survival, there is a culture-nature co-relation that is reinforced regularly by the community. Rituals and festivals performed at various times of the year bring the society together (Bey 2009). This may also mirror the kinship bonds and implications for community ownership of land (Bhattacharjee 2021). As for the sacred groves, the Karbis ensure that particular geographical areas are demarcated as places of religious significance (Timung 2003). Like most tribal societies, the Karbis believed that the deities owned specific territories which are also their dwelling abodes. Each presiding deity *(Sarpo)* of an area is propitiated in the case of clearing of specific patches of land (Phangcho 2001). The items offered include areca-nut, betel leaves, or/and animals such as hen, goat etc. The sacred groves are a living embodiment of the Karbi culture, and they are a repository of traditional resources for the Karbi religion and culture (Teron 2008). Several such floral resources were inseparable from their animistic rituals and indigenous knowledge of health care practices. The Karbis have been preserving and sustaining the resources through their cultural practices which is handed down from generation to generation.

Sacred Groves of the Karbis

Sacred groves are also sites of ancestor veneration worshipped through a chain of rituals. The traditional priests (*kurusar*) officiate the rituals and carry forward the tradition to ensure the sustenance of the indigenous culture. Sometimes the village headman (*sarthe*) of neighbouring villages situated near the particular sacred grove(s) also perform the prayers and rituals in case a specific village/community has been affected by the wrath of the deities in the sacred groves. Each deity of the Karbi religious pantheon has particular jurisdiction over certain aspects of the world and life, and they are worshipped through an elaborate system of offerings. The Karbis believe that the creator of the Universe is *Arnam Songsar* or *Songsar Recho* (7). They also believe in the existence of Heaven and *Barithe* is the God of Heaven. There is no system of idol worship, temples, or shrines as in the Hinduism though annual religious events like *Rongker* are held in a fixed location. The beliefs surrounding sacred groves are not openly discussed, but rather kept in secrecy. The *than* is the abode of the deitie(s), and they are believed to protect the particular territory where they are venerated. There is much awe and fear of the deities and in case of any untoward behavior towards them, it is believed that severe punishment will be incurred on the person in the form of illness and misfortune.

The most well-known and important sacred grove in Karbi Anglong is *Ronghang Rongbong*, located about 16 kms south of Hamren. The significance attached to the sacred site is that it is also a traditional seat of the erstwhile Karbi king, which also gives a sense of the centrality of the ecosystem in Karbi culture. The Karbi king resides here along with other officials. A large wooden pillar is erected in *Ronghang Rongbong*, which is believed to have been brought from the place of origin of the Karbis before they began their migration journey. There are myths associated with human sacrifice being performed at *Ronghang Rongbong* in early days of their history. It is also believed that in the olden days, in a gap of a year or two years, three men used to come to *Ronghang Rongbong* and self-immolate themselves as a sacrifice to the territorial deity to bring peace and well-being to the neighboring villages. The Teron clan is considered a priestly clan in the Karbi society, and they are said to have performed a ceremony where a man was sacrificed yearly by the Teron priest (*Dengjakethe*). The *Ronghang Rongbong* is a sacred site and generally entry is strictly forbidden, although prayers must be offered if a person has to enter (Bhattacharjee 2010). In contemporary times, the sacred grove of *Ronghang Rongbong* houses a complex instituted by the Karbi Anglong Autonomous Council. The titular Karbi king known as *Recho or Lindokpo* is the overarching authority over the political and religious activities of the community. The sacred spatial area of *Ronghang Rongbong* symbolises the legitimate political centre from which is believed to emanate all power.

Another sacred grove is at the *Inglongkiri* hill situated inside the Marat Longri wildlife sanctuary, located 10 kms away from the Diphu subdivision. The local community believes that the abode of the deity should not be disturbed as the deity may unleash wrath on any trespasser. In order to appease the territorial deity, people offer prayers and sacrifices. *Ritasor* is a sacred grove situated in Baithalango of Hamren district. Considered the deity of hills, the deity who resides here is a popular one and worshipped before the start of annual *jhum* cultivation through rituals of animal sacrifice.

Bichikri is considered one of the most critical forests amongst the Karbis. Culturally, it holds an important position as the site is associated with many Karbi traditional beliefs. The grove is protected, and the deity that governs the groves is said to have had an intimate relationship with the Karbis since time immemorial. A community festival is performed by the surrounding villages near the sacred grove known as *Bichikri Rongker*. *Rongker* is an annual religious festival with elaborate rituals performed to appease the territorial deity that governs villages (Bey 2009). Every Karbi village of Karbi Anglong performs *Rongker*. Sacred water bodies are an essential part of sacred groves (Bhattacharjee 2010, 136). There is a belief among the Karbis that *Bichikri* has a sacred connection with a pond which dries up during the monsoon season but swells up during the winter season, in contradistinction to the known seasonal conditions. The pond is also said to be indigenous to certain species of animals and fish that are sacred to the Karbi traditional belief system. As such, fishing in this pond is forbidden and since no fishing activity is performed in this grove, the natural habitat and biodiversity is conserved. It may be noted that this tradition has led to a record-high species-rich flora and fauna amongst all the sacred groves in Assam.

Arlongpuru is a sacred grove located about 4 kms away from Umcheret village, near Hamren. This sacred site is covered with trees, huge rocks and boulders and people go there to worship the rocks. No human activity is permitted in the area except the offering of prayers. An annual ritual to appease the deity is performed by a Karbi priest as the malevolent spirit residing here may bring misfortune to the people if proper rituals are not performed.

There is a *Mahamaya* temple grove situated in Samelangso under Dokmoka block. This sacred site and surrounding forests are protected by the Karbis, and plucking of flowers or plants are strictly forbidden. The Karbis consider *Rek anglong* as the protector of the Karbis and occupies the most important position in the Karbi pantheon. It is a territorial deity and this sacred grove is located on top of a hill. It is said that the deity does not tolerate any polluting activity. This deity is also known for his wrath and there are strict guidelines for the rituals. When Karbis go to this sacred grove to offer prayers, the devotees are forbidden to move around and they are to sit on a flat rock which is located there. traditional priest performs rituals for the people. The rituals are allowed to be performed only by the priest.

Gender and Sacred Space

Women in Karbi society share a symbiotic relationship with forests as it provides them with their livelihood. An important contribution of the women to Karbi economy and cultural heritage is their vast knowledge of forest-based resources (Hansepi and Laisram 2022). Yet, there are gender restrictions imposed on the women regarding entry into the sacred groves. The women are forbidden to enter the designated places in *Ronghang Rongbong* as it is said that they could pollute the environment. This is particularly strict for menstruating women. As such, the Karbi women enter the forest areas for collection of wild fruits and seeds, and on such occasions, they invoke the deities through prayers. The women are careful to refrain from collecting fresh produce of the forest as it is believed to be the property of the deity concerned. They only collect dried twigs and fallen branches and never cut down the trees. The observance of taboos runs deep in many of their practices and this has helped in maintaining natural waterbodies, they make sure that they do not pollute it by throwing waste etc. They draw water without making noise as they believe the waterbodies are home to a water spirit (*langjangsarpi*). There is a dress code for entering the sacred groves and women are generally required to wear clothing that covers the whole body. While the imposition of traditional norms helps to protect the natural resources, it reflects the gender dimension inherent in the society.

Community and Conservation

There are some trees which are grown in the forests and used for making community totems like *Bengvoi Ke-Er* (*Wrightia Coccinea Sims*). In Karbi society, the tree holds a special cultural significance as it represents the five Karbi clans, the Karbi king and the dignitaries (8). Most Karbi prohibitions, bans, and taboos are related to the age-old beliefs in *Honghari.* The Karbis also believed that their well-being and prosperity depends on following the social norms related to taboos. It may be mentioned that taboos are the unwritten, orally transmitted traditions or social rules regulating human behaviour (Colding and Folke 1997). Among many indigenous communities, this practise has been an agency for traditionally protecting and preserving the ecosystem. In case the person falls ill or misfortunes occur, it is believed that the person has unleased the deity's wrath, and a priest (*kurusa*r) is then required to perform rituals to bring back the person to normalcy.

It is widely recognised that the livelihood of the tribal people in India are largely dependent on forest produce and natural resources. For the Karbis, the intricate linkage between nature and culture is rooted in their cultural ethos. The sacred groves are repositories of traditional knowledge such as ethnomedicinal herbs. Each community prescribes its own values, customs, and norms which are synchronized with the worship of spirits that reside in sacred forests. The rituals performed for various territorial deities is a constant reminder that the ecosystem has to be preserved if the society has to sustain for future generations. It is crucial to appease the malevolent spirits to ward off evil. On the other hand, it is believed that the benevolent spirits may sometimes gift the worshippers with elephant tusks, brass materials, paddy, etc. The Karbi society has laid down certain norms in the context of entering the sacred groves. No individual is allowed to carry knives (*tari*) or machetes (*nopak*). It is forbidden to collect dried leaves, twigs, flowers, and fruits from the sacred sites. Since it is believed that the deity has the sense to detect the intentions of the intruder, the Karbis are very careful to follow the prescribed social norms. If the deity senses negative intention, the deity could punish the intruder through mental and physical harm. Some of the common illnesses that an individual may contract include stomach-ache, vertigo, vomiting, etc.

The sacred groves should be visited generally during the daytime, and no human activity and intrusion of any sort are allowed at night. It is believed that the deity is more alert at night and can detect any activity with severe repercussions for the intruder. There are strict codes of conduct whereby they should not utter curse words (*lamjuno*) in any forest. To maintain sanctity of the site, the people have to maintain silence. It is forbidden to capture any insect or animal and even climbing trees are strictly forbidden. No agricultural activities are permitted near the vicinity of the sacred groves. The Karbis believe that *bongkrui* (leopards) and *teke* (tigers) guard the sacred groves, and that they might ferociously attack intruders.

Conclusion

In many ways, religion and culture of the Karbis have been the basis of sustenance of the sacred natural sites in undivided Karbi Anglong district of Assam. They nurtured the forests and conserved its biodiversity through their living tradition carried out through religio-cultural activities. The sacred grove is a time-tested successful model of forest-centered conservations. It cannot be only viewed as biodiversity hotspots as they have a significant role in the social contexts of the indigenous communities who conserve them. The key motivators for people to conserve them lie in their traditional, religious, and social beliefs. The key holders of such beliefs are the elderly, women, and people who still follow the traditional animistic beliefs. However, the communities that have preserved them for generations have shown concern regarding a looming threat towards sacred groves- through illegal encroachment of forests area, deforestation, hunting, and other developmental activities. Informal conservation practices by the local indigenous communities are crucial for biodiversity conservation, especially in an era of conflict, and exploitation of natural resources is at an all-time high (Malik et al. 2016). It is crucial to consider the cultural values and integrity of the local community that has for generations preserved it and ways to uphold and support such cultural values that can effectively enable forest conservation. A good conservation strategy to adopt is to learn the traditional knowledge of the conservation of sacred groves and empower the local community to revitalize their customs, values, and taboos through socioeconomic modernization.

Notes

[1] The term 'biodiversity hotspot' was coined by Norman Meyers in 1988. According to Conservation International (https://www.conservation.org), at present, there are 36 biodiversity hotspots, representing just 2.5% of earth's land surface. The four recognized biodiversity hotspots in India are: the Himalayas, the Indo-Burma region, Sundarbands, and Western Ghats.

[2] The Karbi Anglong (undivided) district, formerly known as Mikir Hills, is the largest of the 31 districts of Assam and was created on 2 February 1970. It is administered by Karbi Anglong Autonomous Council according to the Sixth Schedule of the Constitution of India and has its headquarters at Diphu. It is bounded by Golaghat district, Meghalaya state and Morigaon district in the west, Nagaon and Golaghat districts in the north, and Dima Hasao district and Nagaland state in the south. Prior to this, it was part of a district known as United Mikir and North Cachar Hills District, which bifurcated into two separate districts in 2016 under the banner of Mikir Hills and North Cachar Hills district in the year 1970. In 2016, the West Karbi Anglong was created from the undivided Karbi Anglong district with Hamren as headquarters. See Timung (2003), Das (2006), Bey (2009).

[3] The Karbis are one of the largest ethnic groups in Assam and rank third, next to the Bodos and the Mishings. See Bey (2005), Bordoloi (1987), Phangcho (2001), Terang (2003).

[4] Agriculture required the clearance of large tracts of forestlands for cultivation, which necessitated the role of preserving specific patches of natural vegetation. Authors such as Kosambi (1962), and Pungetti et al. (2012) have noted that sacred groves in India have existed since time immemorial and that it is common to both Hinduism as well as that of animistic beliefs.
[5] Karbis had their traditional belief system and were animists, though a section of them today have been influenced by Hinduism. Teron (2008) has argued that this has seriously affected the religious identity of the Karbis. The conflicted religious identity has been discussed as one of the major reasons for mass scale conversion of Karbis to other religions. See Bhattacharjee (2021), Bey (2009).
[6] *Census of India: 2011*. The district has a sex ratio of 951:1000 [female: male].
[7] *Songsar* seems to be a corrupted version of the Sanskrit term *sansar* which also mean 'universe.' This denotes that the Karbis have borrowed certain beliefs and practices from Hinduism. Since Ahoms were the only Hindus, they have come in contact with, this suggests the impact of Ahoms over the Karbis. See Das (2011), Lyall (2011).
[8] A totem is defined as a plant, animal, bird or any other living thing or material object which a group of people (especially a clan or ethnic affiliation) might consider as sacred as it might have provided them some assistance that ensured it survivals. The contribution of the totem is recognised and revered, and it is forbidden to kill or destroy any totemic plant, animal, bird or object as it is sacrosanct (Adom 2016).

References

Adom, D. 2016. "Asante Indigenous Knowledge Systems: Repositories of Conservation Ethics for Ghana's Biodiversity." In: *Sub-Sahara Indigenous Knowledge and New Discoveries in the New Millenium: Interdisciplinary Approach* 7(2). Sokoto: Uthman Danfodio University.

Anderson, D.G. and E. Berglund. Eds. 2003. *Ethnographies of Conservation: Environmentalism and the Distribution of Privilege.* New York: Berghahn Books.

Berkes, F. 2007. Community-based conservation in a globalized world. *Proceedings of the National Academy of Sciences* 104 (39): 15188-15193.

Bey, Mondolsing. 2005. *A Glimpse of the Socio-political Evolution of the Karbis of Karbi Anglong of Assam*. Diphu: Monjir Press.

Bey, Mondolsing. 2009. *The Karbi Belief.* Diphu: Karbi Lammet Amei.

Bhattacharjee, Somenath. 2010. "Sacred Groves in Karbi Anglong: An Anthropological Observation." *The Eastern Anthropologist* 68(1): 131-142. Accessed May 16, 2020. https://www.researchgate.net/publication/283429009_Sacred_Groves_In_Karbi_Anglong_an_Anthropological_Observation.

Bhattacharjee, S. 2021. "Indigenous Knowledge and Practices Related to the Reproductive Mother-Child Health Issues of the Karbis in Karbi Anglong, Assam." *Ethno Medicine,* 15(1-2): 13-29. Accessed May 18, 2020. DOI: 10.31901/24566772.2021/15.1-2.621.

Burner, A.G., R.E. Gullison, R.E. Rice, and G.A.B. da Fonseca. 2001. "Effectiveness of Parks in Protecting Tropical Biodiversity." *Science* 291: 125-128. Accessed November 8, 2020. DOI: http://dx.doi.org/10.1126/science.291.5501.125.

Chakrabarti, Kunal. 1998. "Recent Approaches to the History of Religion." In: *Recent Perspectives of Early Indian History*, Edited by Romila Thapar. Mumbai: Popular Prakashan Pvt. Ltd.

Colding, Johan, and Carl Folke. 2001. "Social Taboos: 'Invisible; Systems of Local Resource Management and Biological Conservation." *Ecological Applications* 11(2): 584-600.

Das, Girindra Nath. 2006. *Tribal Tradition and Development in the Hill Areas of Assam.* Guwahati: Assam Institute of Research for Tribals and Scheduled Castes. Accessed June 6, 2020. DOI: https://doi.org/10.1890/1051-0761(2001)011 [0584: STISOL]2.0.CO;2.

Das, Indranoshree. 2011. *Folk Medicine of the Karbis of Assam*. Guwahati: Directorate of Assam Institute of Research for Tribals and Scheduled Castes.

Devi, Ashalata, Awadhesh Kumar, and M.L. Khan. 2017. *Final Technical Report on: Sacred Groves of Assam, Biodiversity Status and Strategies for their Conservation*. Napam, Assam: Department of Environmental Science, Tezpur University.

Dudley, Nigel. 2008. *Guidelines for Applying Protected Area Management Categories.* Gland, Switzerland: International Union for Conservation of Nature (IUCN).

Gadgil, Madhav and Vaman D. Vartak. 1976. "The Sacred Groves of Western Ghats in India." *Economic Botany.* 30(2): 152-160. Accessed May 28, 2020. https://www.jstor.org/stable/4253716.

"Global Biodiversity Hotspots with Special Emphasis on Indian Hotspots." Accessed December 17, 2022. http://www.bsienvis.nic.in/database/biodiversity-hotspots-in-india_20500.aspx.

Gluckman, Max. 1965. *Politics, Law and Ritual in Tribal Society*. Chicago: Aldine Publishing Company.

Hansepi, Laxmi and Rena Laisram. 2022. "Karbi Women and Environmental Conservation." *Grassroots Journal of Natural Resources,* 5(3): 1-17. Accessed November 23, 2022.

DOI: https://doi.org/10.33002/nr2581.6853.050301.
Hughes, D.J. and M.D. Chandran Subash. 1997. *Sacred Groves around the Earth: An Overview of Workshop on the Role of Sacred Groves in Conservation and Management of Biological Resources.* Peechi: Kerala Forest Research Institute.
Hughes, J. Donald, and Jim Swan. 1986. "How Much of the Earth is Sacred Space?" *Environmental Review: ER* 10(4): 247–59. Accessed June 17, 2020. DOI: https://doi.org/10.2307/3984349.
Karbi Anglong: Moving Ahead for Biodiversity Conservation. 2010. Guwahati: Assam State Biodiversity Board. Accessed December 18, 2022. https://asbb.assam.gov.in/sites/default/files/swf_utility_folder/departments/asbb_lipl_in_oid_7/portlet/level_2/Booklet_%20Karbi%20Anglong.pdf
Kosambi, D.D. 1962. *Myth and Reality*. Bombay: Popular Press.
Kandari, L.S., V.K. Bisht, M. Bhardwaj, and A.K. Thakur. 2014. "Conservation and Management of Sacred Groves, Myths and Beliefs of Tribal Communities: A Case Study from North-India," *Environmental Systems Research* 3(1). Accessed June 11, 2020. DOI: https://doi.org/10.1186/s40068-014-0016-8.
Lyall, Charles. 2011. [Reprint]. *The Mikirs*. Guwahati: Spectrum Publications.
Mahanta, Pradip J. and Debarshi Nath. 2013. *Cultural Studies: Perspectives from North East India*. Tezpur: Department of Cultural Studies, Tezpur University.
Malhotra, Kailash C., Yogesh Gokhale, Sudipto Chatterjee, and Sanjeev Srivastava. 2001. *Cultural and Ecological Dimensions of Sacred Groves in India*. New Delhi: Indian National Science Academy.
Malik, Z.A., R. Pandey, and A.B. Bhatt. 2016. "Anthropogenic Disturbances and their Impact on Vegetation in Western Himalaya, India," *Journal of Mountain Science* 13(1): 69–82. Accessed June 5, 2020. DOI: https://doi.org/10.1007/s11629-015-3533-7.
Myers, N., R. Mittermeier, and C. Mittermeier, G.A. da Fonseca, and J. Kent. 2000. "Biodiversity Hotspots for Conservation Priorities." *Nature* 403: 853–858. Accessed July 2, 2020. DOI: https://doi.org/10.1038/35002501.
Negi, Chandra S. 2005. "Religion and Biodiversity Conservation: Not a Mere Analogy." *International Journal of Biodiversity Science and Management* 1(2): 85-96. Accessed April 18, 2020. DOI: 10.1080/17451590509618083.

Pakem, B. Ed. 1990. *Nationality, Ethnicity and Cultural Identity in North-East India*. Guwahati: Omsons Publications.

Pal, Tapas and Malay Mukhopadhyay. 2011. "Change in Socio-cultural Dimensions and its Impact on Existing Sacred Grove." *International Journal of Human Social Science* 1(10): 242–250. Accessed April 21, 2020. https://www.semanticscholar.org/paper/Change-in-Socio-Cultural-Dimensions-and-its-Impact-Pal-Net-Jrf/8e005f1242bb438b014989928faaa54fed591e16.

Pathak, N., S. Bhatt, T. Balasinorwala, A. Kothari, and G. Borrini-Feyerabend. 2004. *Community Conserved Areas: A Bold Frontier for Conservation.* Tehran: TILCEPA/IUCN, CENESTA, CMWG and WAMIP.

Pereira, Melvil et al. Ed. 2017. *Exploring Differences, Recording Change: The Karbis of Assam*. Guwahati: North Eastern Social Research Centre.

Phangcho, P.C. 2001. *Karbi Anglong and North-Cachar Hills: A Study of Geography and Culture*. Diphu: Karbi Lammet Amei.

Pungetti, G., G. Oviedo, and D. Hooke. 2012. *Sacred Species and Sites: Advances in Biocultural Conservation.* Cambridge: Cambridge University Press.

Putney, A. 2005. "Building Cultural Support for Protected Areas through Sacred Natural Sites." In: J. McNeely. *Friends for Life*. Gland, Switzerland and Cambridge: International Union for Conservation of Nature (IUCN).

Ray, R. and T.V. Ramachandra. 2010. "Small Sacred Groves in Local Landscape: Are they Really Worthy for Conservation?" *Current Science* 98(9): 1178-1180. Accessed June 19, 2020. https://www.researchgate.net/publication/256936123_Small_sacred_groves_in_local_landscape_are_they_really_worthy_for_conservation.

Ray, R., M.D. Subhash Chandran, and T.V. Ramachandra. nd. *Sacred Grove: Nature Conservation Tradition of the Ancient World.* Bengalore: Centre for Ecological Sciences, IIS. Accessed December 20, 2022. https://wgbis.ces.iisc.ernet.in/biodiversity/sahyadri_enews/newsletter/issue32/article1.htm.

Roy Burman, J.J. 1998. "Syncretic Sacred Groves and Forests." *Eastern Anthropologist* 51(4): 373-381.

Reinach, Salomon. 2003. *Cults, Myths and Religion*. London: Kesinger Publishing.

Terang, Barelong. 2003. *History of the Karbis*. Guwahati: Bharat Offset.

Teron, Dharamsing. Ed. 2008. *Karbi Studies: Memories, Myths and Metaphors.* Guwahati: North-East Printers.

Timung, Lunse. 2003. *Karbi Culture and History*. Diphu: Karbi Lammet Amei.

Tylor, Edward Burnett. 1871. *Primitive Culture: Researches into the Development of Mythology, Philosophy, Religion, Art,* and *Custom*. Michigan: University of Michigan.

Verschuuren, Bas, Robert Wild, Jeffrey McNeely, and Gonzalo Oviedo. Eds. 2010. *Sacred Natural Sites: Conserving Nature and Culture.* London: Earthscan.

"Why are Biodiversity Hotspots Important?" Accessed December 15, 2022. https://www.conservation.org/priorities/biodiversity-hotspots.

Wild, Robert and Christopher McLeod. Eds. 2008. *Sacred Natural Sites: Guidelines for Protected Area Managers*. Gland, Switzerland: International Union for Conservation of Nature (IUCN).

CHAPTER II

RITUAL PRACTICES AND NATURE CONSERVATION AMONG THE MISINGS OF ASSAM

RIPUN DOLEY

The Misings (1) are the second largest tribal group in Assam, only next to the Bodos, with a population of 6,80,424 individuals (*Census of India 2011*). Linguistically Tibeto-Burman, (Grierson 1927, 40) they belong to the Mongoloid stock and at present inhabit the ten districts of Upper Assam *viz*; Dhemaji, Lakhimpur, Biswanath, Majuli, Jorhat, Sibasagar, Golaghat, Tinsukia, Dibrugarh and Sonitpur. They have mainly settled along the banks of the rivers Brahmaputra, Subansiri, Simen, Burusuti, Gunasuti, Tongani, Jiadhol, Boginodi, Ranganodi, Dikrong, Pichola, Burai and Jiabharali, Dibru, lower Dihing, Dishang, Dikhow and Dhansiri. Prior to their migration to the plains of Assam, the Misings mainly practised hunting-gathering and shifting (*jhum*) cultivation. They learnt the technique of cultivation after their settlement in the plains of Assam (Padun 1989, 15). It may be noted that Assam, situated Northeast India (2), known for its diverse culture and each culture has its own set of tradition and method of conserving nature which are generally associated with their religious beliefs and taboos.

Sacred groves have played a vital role in the protection and preservation of the ecosystem worldwide. A rich biodiversity hotspot, these virgin patches of forests are protected by the local people due to their cultural and religious beliefs (Khan et al. 2008, 278). In India as well as other parts of the world, sacred groves are found mainly in tribal dominated areas bearing different names (Bhakat and Pandit 2004, 37-43). The largest known documented sacred groves in India are Hariyali (Uttarakhand), and Shipin (Himachal Pradesh). The natural sites are also associated with myths and taboos of specific communities which ensures the preservation and protection of the sacred landscape. In traditional societies, the lives of the people are governed by social norms and customs; and there is a fear that any disrespect shown to the tradition will harm the individual or community through the curse of premature death, sickness, poverty etc. (Sangma 2006, 56). The centrality of rituals in reinforcing the sustenance of the sacred groves includes taboos which forbid certain social customs and practices.

Objective and Methodology

This study examines the traditional religious belief system of the Misings as agency for preservation and management of natural resources. It will gain insights into the ways in which ritual as a component of Mising culture have played a crucial role in nature conservation. The importance of rituals in religious activities have been explored by anthropologists such as Victor Turner, who defined 'ritual' to mean 'prescribed formal behaviour for occasions not given over to technological routine, having references to belief in mythical beings or powers' (Turner 1967, 19). Prescribed ways of carrying on religious acts and procedures, ways of worshipping, praying, chanting, sacrificing, making offerings etc., are practiced in rituals. Ritual is an inevitable component of culture, extending from the large-scale social and political processes to the most intimate aspects of self-experience (Kluckhohn 1958, 146). As such, every society follows social norms which can be divided into four categories, viz., beliefs, rituals, superstitions, and taboos. Taboo is a ritual prohibition that is rooted in specific cultures. MacDonald (1977, 735) opines that historically taboos have been used to impose restrictions for social wellbeing, and their legacy is passed on from one generation to another.

The Mising Worldview

An ancestor worshipping community, the Misings of Assam perform various rites and rituals in their everyday lives which include worship of both benevolent and malevolent spirits. The religious philosophy of the Mising is based on animism and they worship the forces of nature such as the sun, the moon, and rain etc. It may be mentioned that the Misings in their early history of migration were hill-dwellers, and their lives have been shaped by environmental factors, such as hills, forests, rivers, wild animals, rains, earthquakes, disease, epidemic etc. In fact, the religious beliefs and practices of the community may be said to be a reflection of the awe and reverence of the natural forces which affect their daily lives. The Mising worldview consists of three major aspects, *viz*; the creation of the universe (*keyum*), existence of spirits (*uie)* and that of the human soul (*ya:loo*). They believe that there are three worlds in the universe which include the *regi-regam* (heaven or kingdom of gods), the *dong-a:mong* (earth), and *uie-a:mong* (abode of spirits). There are myths which attempt to explain the significance of heaven and earth and how they exist within the orbit of the solar system. The abode of the spirits is believed to exist outside the solar system and hence remain in perpetual darkness (Padung 1989, 384).

Like most traditional societies, the Misings also believe in a Creation Myth, which is preserved as an oral tradition and known as *Mibu Aa:bangs* (Regon 1989, 366). Cosmogony myths are prevalent among many cultures around the globe such as in Africa, the Ainu of the Northern Japanese island, the Central Australians etc. (Gwinn 1988, 368-370). For the Misings, the creators of the universe are *Se:di-Babu* (Se:di, the father) and *Me:lo-Nane* (Me:lo, the mother). They are believed to have evolved from *Keyum> Yumkang> Ka:si> Siyan> Anbo> Bomug> Mukseng> Se:di* (Taid 2008, 57). According to the myths *Se:di-Babu* and *Me:lo-Nane* created all animate and inanimate objects. It is said that *Se:di-Me:lo* gave birth to *Diling> Litung>Tuye> Yepe> Pedong* (rain). *Pedong* is the female principle that gave birth to both human beings and the spirits (Taid 2008, 57). The Misings invoke the name of their Ultimate Parents, namely, *Do:nyi Ane* (Sun Mother) and *Po:lo Abu* (Moon Father) on all important occasions apart from *Se:di-Me:lo*. It is believed that *Do:nyi-Po:lo* sprang directly from *Se:di-Me:lo. Do:nyi-Po:lo* literally meaning 'Sun-Moon'. It is important to note that they do not worship *Se:di-Me:lo* and *Do:nyi-Po:lo* but instead of that they take oath and invoke their name in every rituals as they are believed to be the custodian of law and truth. Like other primitive society, the Misings have also unique myth about the creation of the universe and their repositories of Creation Myth are found in the religious verses that preserved as oral tradition and that is known as known as *Mibu Aa:bangs* (Regon 1989, 366).

Ritual Specialists and Sacred Ecology

The traditional priests who preside over rituals are called *Mibu (mi:*man; *bu:* father, meaning father of man) (Regan 1989, 367). There is a myth regarding the origin of the *Mibu*. The narration is about *Abotani,* (*abo*: father; *tani*: man, meaning father of man) whom the Misings believe to be the first man of their community, and who married Yasi, the daughter of *Do:nyi* (Sun). Due to a severe famine on the earth, *Abotani* is believed to have gone to the house of *Do:nyi* seeking solution from such deadly woes.

Before reaching the house of *Do:nyi*, he entered the house of *Ko:je-Yango* (goddess of wealth and fertility). Hearing the woes of mankind *Ko:je-Yango* offered some paddy seeds and taught him how to sow while also warning him not to visit the paddy field before five days, after sowing the seeds. However, he forgot the warning and visited the paddy field before the appointed time. There, he saw that a dove had eaten all the seeds. Soon *Abotani* realized his fault and started crying aloud. Hearing his cry, *Seukari* (God of Hunting) felt sad for him and killed the dove by an arrow and advised *Abotani* to collect the remaining seeds from the throat of the dove for further cultivation. However, *Seukari* put a condition that in lieu of his favour, *Abotani* would have to arrange a ceremony to appease *Seukari* and *Ko:je-Yango* after the harvesting. *Abotani* arranged a great ceremony where *Seukari* himself participated in it and also selected a competent person from the gathering who could act as a priest. This priest is known as *Yumging,* and later also as *Mibu* (Regan 1989, 368-369*)*.

The *Mibu* who is considered to possess some supernatural powers and he plays the role mediator between the world of the spirits and the humans. He is kept busy throughout the year conducting various rites and ceremonies at various levels such as in the home, in the locality or the village. It is said that the *ab:angs* (religious hymns) to be recited during the ceremony is taught to him by a forest spirit called *upom uies*. The training process to become a priest is known as *mibu a:regnam* (Pegu 1989, 339-340). They also give him a sword called *Yoksa*. The role of the *Mibu* in Mising society is significant as it is to him the members of the community turn for help in case they are afflicted with disease or disaster. Apart from his priestly duties, he also plays the role of a healer or medicine-man (Pegu 2011, 111*)*. In an attempt to help the victims of misfortunes, the *Mibu* recites the *aa:bangs* and claim that he communicated with the spirits who are believed to have caused the problem. The process is called *Mibu Dag:nam* and is performed only at night time. The *Mibu* invites the victim family to secretly place a small packet of rice wrapped in *taling a:ne (alpinia taro)* in the ceiling of the room where the *Mibu* is supposed to sleep at night. If the *Mibu* is bestowed with the spiritual power, he was to be able to tell the details of the family background as well as the causes of the problem and its solution.

According to a myth, the *upom uies* (benevolent spirits)*,* who teaches *aa:bang* (hymn) to the *Mibu* dwells on *peepal* trees (*ficus religiosa*) in dense forests. The *upom uie* are generally considered to be harmless and benevolent; but become ferocious if the trees on which they reside are disturbed by the act of cutting it (Pegu 1989, 339-340). Therefore, it is a taboo to plant *peepal* tree in front of a Mising house or cut the tree which is the dwelling place of the spirits. If there is need to cut the *peepal* tree, then there is a ritual they must offer *a:pong* (rice beer) and *tamul-pan* (betel nut) to the spirits and then after seeking permission from the spirits they dare to cut the tree (Doley 2022). Apart from it, the Misings do not cut devil's tree (*alstonia scholaris*) and they do not use the wood of the devil's tree for any purpose in their house as they believe that the spirits dwell on such trees. The *taling (alpinia taro)* is considered as a sacred plant by the Misings and the leaves are used for rituals (Doley 2022) There is a tradition to prepare *pu:rang a:pin* (boiled rice) wrapped with *taling a:ne* to offer the ancestors on the occasion of *Ali-aye-Ligang and Po:rag* festival. Apart from this, the leaf is also used in other traditional rituals relating to spirit appeasement. This plant is grown in every household due to its requirement in the Mising ritual ceremonies.

Spiritual Dimension of Nature

The worship of spirits, both benevolent and malevolent is central to the belief system of the Misings. In fact, the spirits inhabit all of nature. Thus, the malevolent spirits represent the destructive forces of nature such as storm, thunder, lightning, and cyclone and are known by names such as *Taleng Uies, Urom Po:sum* and *Epom-Yapom* etc. They cause disease, physical disabilities, accidents, unnatural death, epidemics. The *Taleng Uie* (*taleng:* heaven, sky*; uie:* spirit, god) used to be appeased through ritual performances once in five years, although in recent years, this is not strictly followed. The *Urom Po:sum* are the spirits of the deceased who died of unnatural causes. The *Epom-Yapom* are the forest spirits residing in trees and capable of causing harm to humans. The Misings also seek blessings and protection from the benevolent spirits through specific rituals performed at particular times of the year (Pegu 2011, 111). The *Zakura Taleng Uie* is a very important spirit and elaborate rituals are performed outside the house. A temporary house is made by bamboo and banana leaf. The items required for performing the ritual include a black gilt for mother earth, five hens and five roosters in different colours, two types of rice beer *viz*; *nogin* and *poro a: pong,* banana, rice flour, fruits, and leafy vegetables (Doley 2022). The traditional priest officiates the rituals. To the east of the temporary-built house, a medium sized branch of *semal* tree (*bombax*) is planted and few drops of *poro a: pong* (rice beer) and blood is sprayed on it. Though they do not worship the *semal* tree, it is considered sacred. Therefore, it is taboo to cut the tree as they believe that spirits live there. Another sacred plant is the *piro* or wild reed which is used in many auspicious occasions such as *Ali-aye-Ligang* and *Po:rag* festivals. The wild reed is also used at the ladder in a Mising house at the time of welcoming a bride for the first time. The *Kencha Taleng uie* is a heavenly spirit which the Mising community invoke when a person loses the eyesight etc. It is mandatory to perform the ritual for appeasement in a site far away from the village, particularly on the banks of a river bank.

The *Yumrang Uie* (*yumrang*: forest; *uie*: spirit) is another malevolent spirit in the traditional Misings religious belief system. As the Misings settled on the banks of the rivers, there was a close connection with nature which included forests. It was important for them to live in harmony with the surroundings for which rituals were central to their daily existence. The two sacred trees which are used in rituals are *peepal* (*ficus religiosa*) and *semal* tree *(bombax*). The forest spirits to be appeased included *Pumsu uie, Sikari uie* and *Kencha Dangaria* or *Dhal Bhagna*. Generally, when a person suffers after being attacked or threatened in the forest by the forest spirits in the form of any wild beasts, a ritual for these spirits is performed under the *peepal* or *seemal* tree, far away from the house. In case there are no trees of these two specific species available in the forest, it is planted for the forest spirits. The crucial point is the belief that the trees are the abode of the spirit. The *Aag:nam* ritual is performed a few days prior to this in which an iron rod is dedicated in the name of the forest spirits. In the *Pumsu uie* two red colour roosters and three hens are required along with rice beer and other items. On the other hand, in *Sikari Uie* and *Kencha Dangaria* or *Dhal Bhagna uie* only one white colour and red colour rooster are required respectively. It may be noted that the forest spirits are said to be more active at night and so one should avoid visiting the forest at night time. In general, there are many restrictions for entering the sacred site.

Water is an essential item for survival and the Misings pay respect to *Asi Uie* (*asi*: water; *uie*: spirit). This malevolent spirit is much feared as it is said to have the power to kill a person in the event of it becoming angry. The Misings believe that spirits engage in fishing activities at night and therefore the humans should not disturb them during that time. If for any reason people go fishing at night, they must pray to their ancestors for their safe return. The ritual to appease the spirit and it is performed by the riverside. The ritual for *Asi Uie* is performed by the riverside.

Kum:Sung Uie (*kum:sung:* granary; *uie:* spirit) is the granary spirit and is performed to seek for good harvest. The *Kum:sung uie* can be divided into two categories viz. *Lakhimi Uie* and *Potharua Dhangaria*. The *Lakhimi uie* is more elaborate and expensive than the *Potharua Dhangaria* and is performed below the granary (traditional granary is raised from the ground).

According to mythology, *Ko:je-yango* (Lakshmi Devi) gave few seeds to Abotani (first man of the Tani group) and also taught him how to sow it and in honour of *Ko:je-yango.* The *Ya:lo Gog:nam* (calling of shadow) of *Ko:jeyango* is important ritual of *Lakhimi uie*. It is believed that the shadow of *Ko:je-yango* is hiding in the form of fish somewhere in the water due to resentment to the family. So, in the early morning they come to the river bank and try to catch the fish from water by using *jakoi* (a triangular bamboo scoop for fishing). The *Potharua Dhangaria* is performed in the agricultural field. It requires one red colour rooster and hen. The *Mibu* as ritual specialist officiates in the rituals and ensures the continuity of the tradition which displays indigenous wisdom though use of specific flora and fauna.

The Misings believe that the *ya:lo* or shadow accompanies a person wherever he or she goes and permanent separation of the shadow from the body implies death of the person. The shadow may leave an individual temporarily during sleep or in sickness. The *Mibu* can bring back a *ya:lo* by performing a ritual at night called *Mibu-Dag:nam.* According to some myths, the priest takes a journey by taking the form of a bird, tiger or snake in order to cross rivers, mountains and forests to bring back the particular *ya:lo* (Padun 1989, 386). There is a tussle between the spirits and the priest and thus in order to appease the spirits, the items offered include a pig, hen or rooster. There is a merry-making ritual in which the priest invites the spirit through his *a:bang* and the sick person whose shadow is being sought from the spirits is allowed to eat the cooked items. The spirit releases the *yo:lo* of the sick person only after this ritual.

Do:bur is a propitiation ritual performed for Mother Earth by the Misings in the event of a misfortune that occurred in the village. The aim of the ritual is to seek blessings for the welfare of the entire community (Padun 1989, 415). There are four types of *Do:bur* rituals viz., *Mo:pun Do:bur, Raban Ko:son, Burte' Do:bur* and *Ko:son Do:bur.*

The Misings primary depend for their livelihood on agriculture and, therefore, most of their rituals are related with agricultural activities. *Mo:pun Do:bur* (*Mo*: stands soil and *pun* stands flower, meaning soil flower) is performed when there is destruction or damage of their crops. The ritual is performed in a paddy field by sacrificing a black colour gilt and two or three hens, praying to *Ko:je -Yango*, (the Goddess of wealth and fertility) and other supernatural forces for protection and good harvest of their paddy field. *Raban Ko:son Do:bur* is performed to prevent suffering of the community. They offer the blood of a gilt or a hen to supernatural forces. The *Burte Do:bur* is performed at the individual, family or village level to ward off evil spirits that may harm them. In the morning of the ritual day, the male members of the village visit every house and beat the wall of each house with stick and wooden rods to drive away evil spirits called *Ajeng Dom:nam* (*Chang Kubua*) by chanting loudly, as thus:

'ajen:gn gu:dung ho
Do:nyi-Po:lo Se:di-Me:lo no Ta:dlangka
Si:lo Pak:ko tani-takor
so:bo-menjek arig-a:ger
pu:po zar:pola dumo:tehka'

(Hemo Doley 2022)

[Free Translation. 'Witnessing *Do:nyi-Po:lo, Se:di-Me:lo*, today we are coming here to drive away the evil spirits of the house through beating the house and from today we wish wellbeing of the family member as well as their livestock and agricultural field."]

The ritual items are called *ajeng* which includes rice, chicken, *a:pong* (rice beer), chilly, salt and pepper and it is brought to the banks of a river, which is the site of the ritual. Here, two symbolic representations of 'dragon swallowing an egg' are prepared from *piro* and *tabeng* (wild reed with cutting edge) which is decorated with split bamboo. They are placed on the alter facing the rising sun (Pegu 2011, 148-149). The prayers are addressed to the 'powerful forces of nature' such as the moon, sun, *Karsing-Kartadang*, cloud, rain and air. The worshippers implore the gods to accept the offerings and bless the people with health and happiness. After the end of the ritual all participants make a procession marking the spatial coverage of the village symbolising driving away of the malevolent spirits from the entire community.

The Misings follow very strict rules during and after the *Do:bur* ritual. During these days there should be no destruction of nature or harm inflicted to animals. Being exclusive to the village, no member of the community is allowed to move out of the village on the ritual day. In case there is an emergency, the person must move out of the village on the previous day. In case any outsider entered the village, then the person has to stay in the village till the ritual is complete. The tradition of not allowing participation of women and children in the rituals indicate the patriarchal nature of the Mising social structure. As such menstruation and childbirth are also considered inauspicious and the women during such periods are kept out of the village (Taid 2002, 59).

Tradition and Nature Conservation

Ancestor worship is at the core of the religious belief system of the Misings. The departed forefathers are considered as benevolent spirits and they protect the family from the evil spirits. So, they worship and propitiate them at regular interval in order to ensure the well-being of the household. This ritual is performed once in every five years. The *urom-to:sag* and *urom-posum* rituals require sacrifice of boars when the name of the forefather is invoked to seek blessings for the family. There are certain taboos related with this ritual. Pumpkin cannot be used in this ritual. The meat of the boar must be consumed within the family as the *gu:min-so:ings* (ancestor spirits) enter the bodies of the family members and through their bodies the dead ancestors are believed to partake the meat (Doley 2022).

The *Ali-Aye-Ligang* (*ali*: root, seed; *ai*: fruit; *ligang*: sowing) is the most important festival of the Misings held in spring season every year on the first Wednesday of the month of *Ginmur-Polo/ Phalgun* (February-March) (Pegu 2011, 33). There is also a ceremony performed at home in the early morning where the head of each family offer a few drops of rice beer in the four corners of the *meram* or hearth chanting prayers by invoking the name of their forefathers (Mipun 1987, 27). Immediately after this act, he goes to the paddy field carrying with him a handful of seeds, a sword, *a: pong* (rice beer), *pu:rang a:pin* (boiled rice wrapped in *taling* leaves) ginger, *piro* (wild reed) cotton seed etc. The symbolic sowing of paddy seeds is done chanting the name of their forefather seeking their blessings for good harvest, and also pray to Mother Earth to witness the ritual of sowing seeds. The prepared food is offered at first to the forefathers and thereafter to *Donyi-Polo* and *Se:di-Me:lo*. In the evening the Goddess of Fertility *Ko:je-Yango* is offered prayer for good harvest (Regan 1989, 122). There is a feast enjoyed in every household with rice beer, meat, and fish in which the members of the village and outsiders are also welcomed. On the second day, the youth perform the Mising traditional dance and they are offered some traditional food items etc. On the third day, a farewell feast is arranged with the items collected on the second day cultural activities. The fourth day is observed as forbidden day when all members of the community is expected to refrain from work.

In Northeast India, *a:pong* (rice beer) is an integral part of the socio-cultural and religious life. The Misings prepare rice beer using indigenous medicinal plants (Kujur and Kandir 2015, 107). There are two types of rice beer viz, *nogin a: pong* (white in colour) and *po:ro a:pong* (dark greenish colour due to the use of ash of paddy husk and straw). The *nogin a:pong* is used by many ethnic communities in Assam including Bodo, Deuri, Sonowal Kachari, and Tiwa etc. The *po:ro a:pong* is an exclusive product of the Misings inhabiting the plains of Assam. It may be noted that the *a:pong* is considered sacred and is first offered to the ancestor and spirits before use for other purposes. This *a:pong* is a mandatory item in all rituals conducted by the Misings. Traditionally, the *a: pong* is produced by fermenting cooked rice with locally prepared cake called *epop*. It is said that fifty different plants are used to make the mixture of *epop*. Generally, it is made from mixture of rice flour and different plants ingredients (Pegu 2022) These plants were abundantly grown in the forests at the foothills of Northeast as well in the plains of Assam. It may be mentioned that alcohol (or rice beer) is a cultural artifact; the form and meaning of drinking alcoholic beverages are culturally defined, as are the use of any other major artifacts.' As Mandelbaum (1965, 281) explains, 'the act of drinking is part of a larger cultural configuration.

Conclusion

Ritual practices of indigenous communities provide useful information on religious activities linked to nature conservation. The Misings worship the forces of nature throughout the year at the individual, family or village level. Their worldview centres around the Sun and Moon which are linked to their Creation Myth. They propitiate various spirits of the forest, water etc. which ensures their connection with the environmental surroundings on a day-to-day basis. In fact, the religious belief system of the Misings mirror their sacred landscape such as hills, forests, and rivers. In following the taboos related to various aspects of their culture, there is a constant reminder of the need to protect natural resources. The *Mibu* as the ritual specialist mediates between the humans and the spirit world to appease the evil spirits so that they do not cause harm to the people. A key aspect of ritual practices is the centrality of the flora and fauna available to the Misings which are regularly in demand for the occasions. The medicinal plants and sacred groves are associated with the religious beliefs and rituals of the Misings which not only protect the plants but also help in biodiversity conservation.

Notes

[1] The spelling 'Mising' was adopted by the *Mising Agom Kebang* (MAK) or Mising Literary Society in 1973 CE.

[2] Northeast India is largely inhabited by Indo-Mongoloids or Kiratas. The region is a veritable paradise for anthropologists, being home to at least two hundred tribes with a rich cultural heritage. The state of Assam is inhabited by as many as twenty-

three major tribal groups, nine in the plain districts and fourteen in the two hill districts. Unlike other states of the Indian Union, the ethnic communities of Assam are classified into two categories based on the geographical location viz; tribes living in the hill districts and the plains' tribes. The former remained isolated till the advent of the British whereas the latter had come into contact with other non-tribal communities at different periods of history. While there is no denying that there are considerable diversities within the region, yet certain common traits are apparent since the communities presently at different stages of cultural transformation are essentially tribal-based societies.

References

Bhakat, R.K. and P.K. Pandit. 2004. "An Inventory of Medicinal Plant of Some Sacred Groves of Purulia District, West Bengal." *Indian Forester* 130(1): 37-43.

Borah, Dipankar, Puranjoy Mipun, Jintu Sarma, Chiranjib Mili, and Deiji Narah. 2021. "Quantitative Documentation of Traditionally used Medicinal Plants and Their Significance to Healthcare among the Mising community of Northeast India." *Ecological Questions* 32(4) September: 1-62.

Census of India 2011. Assam: Directorate of Census Operations.

Dhar. S. 2014. "Socio-Economic and Demographic Status of Assam: A Comparative Analysis of Assam with India." *International Journal of Humanities and Social Sciences* 1(3): 108–117.

Doley, Hemo. Mising Priest (Interview conducted by the author in Gohpur, Biswanath, Assam, India, on February 15, 2022.

Doley, Sibanath. Retired. Teacher at Gohpur Higher Secondary School. (Interview conducted by the author in Gohpur, Biswanath, Assam, India, on March 2, 2022.

Grierson, G.A. 1927. *Linguistic Survey of India* 1, (Part 1): 40ff. Delhi: Motilal Banarsidass.

Gwinn, Robert P. Eds. 1988. *The New Encyclopaedia Britannica*: *Micropaedia, Knowledge in Depth* 17: 368-370. University of Chicago: Encyclopaedia Britannica Inc.

Khan, M.L. et al. 2008. "The Sacred Groves and their Significance in Conserving Biodiversity: An Overview." *International Journal of Ecology and Environmental Science* 34(3): 278. New Delhi: National Institution of Ecology.

Kluckhohn, Clyde. 1958. "*Myths and Rituals: A General Theory.*" In: William A. Lessa and Evon Z. Vogt. Eds. *Reader in Comparative Religion: An Anthropological Approach*. New York: Row, Peterson and Company.

Kujur, Salomy and Kunul Kandir. 2015. "Studies on Ethnomedicinal Plants Used in Rice-Beer (Handia) by 'Ho' Tribe of Jharkhand." *International Journal Medicinal Plants Used in Rice-beer for Exchange of Knowledge* 2(1): 104-108.

Mandelbaum, David G. 1965. "Alcohol and Culture." *Current Anthropology* 6(3): 281-293. Chicago The University of Chicago.

Mipun, Jatin. 1987. *The Mishings (Miri) of Assam: Development of a New Lifestyle.* New Delhi: Gian Publishing House.

O'Doherty, E.F. 1960. "Taboo, Ritual and Religion." *Studies: An Irish Quarterly Review* 49 (194): 131-143. Accessed December 17, 2022. https://www.jstor.org/stable/30099141.

Pegu, Indeswar. 1989. *"Mising Dharma-Biswas Aru Parompora."* In: *Mising Sanskritir Alekhya*, Edited by Brigumoni Kagyung. Guwahati: Lawyer's Book Stall.

Padun, Nahendra. 1989. "*Mising Sanskriti."* In: *Mising Sanskritir Alekhya*, Edited by Brigumoni Kagyung. Guwahati: Lawyer's Book Stall.

Padung, Nahendra. 1989. "*Yaloo Ba Atmar Punaranayan*." In: *Mising Sanskritir Alekhya*. Edited by Brigumoni Kagyung. Guwahati: Lawyers' Book Stall.

Padun, Nahendra. 1989. "*Do:bur Puja*." In: *Mising Sanskritir Alekhya*, edited by Brigumoni Kagyung Guwahati: Lawyer's Book Stall.

Pegu, Nomal. 2011. [Second Revised Edition]. *The Mishings of the Brahmaputra Valley* Milon Nagar Dibrugarh: Nomal Pegu.

Pegu, Pabitra Kumar. 2011. *Mising Samaj-Sanskritir Rengani.* Dibrugarh: Kaustubh Prakashan.

Pegu, Sharu. Interviewed by the author in Gohpur, Biswanath, Assam, India, on February15, 2022.

Regan, Durga. 1989. "*Ali-aye-Ligang: Eyar Baishisthya."* In: *Mising Sanskritir Alekhya* Edited by Brigumoni Kagyung. Guwahati: Lawyer's Book Stall.

Regan, Durga. 1989. "*Miboo Ba Miroo Aru Miboo Aa:bang"* In: *Mising Sanskritir Alekhya.* Edited by Brigumoni Kagyung. Guwahati: Lawyer's Book Stall.

Rongpi, Roselin. *Culture and Alcoholism: An Analytical Study on Consumption* (Nasha Mukt Bharat Abhiyan, in 272 Districts). Government of India: Ministry of Social Justice and Empowerment.

Sangma, C.T. 2006. *Meghalaya: Yours to Discover.* Guwahati: DVS Publisher.

Taid, Tabu. 2002. "The Religious Culture of the Misings." In: *Traditional Customs and Rituals of North-East India.* Edited by Dr. P. C. Sarma. Guwahati: Vivekananda Kendra Institute of Culture.

Turner, Victor. 1967. *The Forest of Symbols: Aspect of Ndembu Ritual.* New York: Cornell University Press.

CHAPTER III

OF GROVES AND THE GODDESS: WOMEN, ECOLOGY, AND SACREDNESS

SAKTI SEKHAR DASH

Following his exile from Thebes, Oedipus made his way to the city of Colonus with his daughter Antigone. They arrived at a peculiar place on the outskirts of the city, one that Antigone declared to be sacred. It was a wonderous place:

> … flourishing
> with laurel, olive, and vine, every branch
> thick with sweet-singing nightingales
>
> (*Oedipus at Colonus*, 1.16-18).

To rest his tired and frail body, Oedipus sat on a stone, when a stranger arrived and urged them to leave the place. The place was sacred and Oedipus had defiled it by trespassing. But Oedipus declared that he had no intention of leaving the place. From the countryman's account, the place is the sacred abode of the Goddesses who went by the name of the Eumenides. The most wretched of humankind, Oedipus could be cleansed of his sins through sacred agencies. Upon learning that the place is dedicated to the fearful Eumenides, he sought sanctuary in the sacred grove. The grove also acted as his final resting place, once the Eumenides had purged him of his dreadful sins. The grove linked the sacred world of the Eumenides with the profane and mundane world of the humans. A few individuals like Oedipus who were permitted to enter the grove gained the status of demigods; not only were they cleansed, but also initiated into the divine order.

The initiation of Oedipus into the sacred or divine order took place in the grove dedicated to the Eumenides. They were also known as the Furies, goddesses of vengeance and retributive justice. "They were usually said to have been born from the blood of Uranus that fell upon Gaia, the Earth, when Cronos castrated him, i.e., they were chthonic (earth) deities" (Grant and Hazel 2005, 222). In Greek religion and mythology, Gaia was revered as the earth goddess; being one of the first divinities to emerge from Chaos, she was the *magna mater*, mother of all life forms in the creation. In the words of Robin Hard, "Gaia will be the progenitor of all that is positive and substantial in the world, including the features of the physical universe that have yet to emerge, and the deities who preside over every department of nature, and all the great gods and goddesses" (Hard 2004, 24). The cult of Gaia was practiced throughout ancient Greece and through her worship, the earth was venerated as the great nurturing mother. She was described as "the giver of dreams and the nourisher of plants" (Taft 2004, 24). According to the Olympian Creation myth, Gaia's union with Uranus resulted in the birth of all life forms; whereas Gaia symbolised the earth, Uranus represented the sky. As the Great Mother, all life forms, plants, animals, and every aspect of nature were sacred to Gaia. Her worship pre-dated the Hellenic civilization and no temples were erected in her honour. On the contrary, her presence was venerated through natural elements and libations offered to her were poured into a cleft in the earth. These clefts were typically situated in groves sacred to Gaia, where the devoted participated in rituals hosted in her honour.

Gaia was not the only deity venerated as the Great Mother Earth; in Vedic Hinduism *Prithvi* was widely worshipped in India. Like Gaia, Prithvi also emerged at the dawn of creation, and brought forth various life forms. Like her Greek counterpart, she represented the fertility and nurturing aspects of the earth. The exact origins of earth worship in India are obscure; before the advent of the Aryans, native tribes and communities used to venerate the earth as the Mother Goddess. Her rites and rituals were performed in groves, forests, and other places linked to the natural world. The sacred groves where the worship of the earth was carried out were associated with natural fertility. The vegetation of sacred places, including the groves fundamentally influenced the religious outlook of the human communities. Trees have commanded reverence from the humans for thousands of years. Their longevity, medicinal qualities, and nurturing qualities equated them with the divine which accorded them a sacred stature.

The sacred groves usually involved a clump of trees growing around the sacred place, which functioned as the open-air shrine for the different rites and rituals. The place, including the shrine and the trees became a part of the sacred sanctuary. In these sanctuaries, the priests and shamans practiced healing and prophesying. The Delphic oracle dedicated to Apollo was originally the seat of Gaia's worship. In the pre-Hellenic days, the

Omphalos or navel stone was worshipped at Delphi testifying to the prevalence of Gaia's cult. Apart from venerating Gaia as the Mother Goddess, Delphi also acquired renown for the prophecies. Even after the worship of Apollo became dominant at Delphi, the rituals and practice of prophesying didn't change. For the Greeks, Mother Earth Gaia spoke through the Delphic Oracle, and the Omphalos continued to be worshipped inside Apollo's shrine.

For the Greeks, Indians, Africans, Romans, Britons, and other communities, the trees and groves in particular were linked to the fertility of the land. Thus, they became inextricably tied to the fertility cults prevalent across various cultures and religions. The Great Mother's worship by various communities represented the integration between religion and ecological sensibility. In the words of Monica Sjoo and Barbara Mor (1991, 15):

> Isis, Mawu-Lisa, Demeter, Gaia, Shakti, Dakinis, Shekhinah, Astarte, Ishtar, Rhea, Freya, Nerthus, Brigid, Danu—call Her what you may—has been with us from the beginning and awaits us now. She is the beauty of the green earth, the life-giving waters, the consuming fire, the radiant moon, and the fiery sun. She is Star Goddess and Spiderwoman; she weaves the luminous web that creates the universe. As earth, the great planetary Spirit-Being, she germinates life within Her dark womb.

Figurines of the Mother Goddess or the Great Mother have been discovered in various parts of the world, including the Southern Anatolian region, Crete, Mycenae, and South Asia. The earliest figurines date back to roughly 30,000 BCE, while the figurine discovered at Mohenjodaro dates back to 3000 BCE. Before the Great Mother began to be worshipped through images and figurines, she had been venerated through natural symbols. In this context, Erich Neumann (2015, 78) adds:

> Before the comprehensive human *figure* of the Great Mother appeared, innumerable symbols belonging to her still-unformed image arose spontaneously. These symbols—particularly nature symbols from every realm of nature—are in a sense signed with the image of the Great Mother, which, whether they be stone or tree, pool, fruit, or animal, lives in them and is identified with them. Gradually, they become linked with the figure of the Great Mother as attributes and form the wreath of symbols that surrounds the archetypal figure and manifests itself in rite and myth.

The veneration of the Mother Goddess through symbols of nature led to the creation of various cults, myths, and religious practices worshipping various aspects of nature, including plants and trees. Out of this veneration, the concept of sacred groves took root.

Culture, Cult, and Sacred Grove

Across different cultures, trees and plants have been a part of the multitude of natural objects and symbols contemplated by the human community in their efforts to construct a system of faith. The plants, trees, and vegetation not only represented the charms and bounties of the natural world, but also initiated the human communities into the mysteries of creation. The forests and groves harboured a diverse variety of animal life, thereby creating a complex web of life. The abilities of plants and trees to sustain and nurture other life forms equated them with the divine forces revered by the humans. Unsurprisingly, for the communities seeking to venerate the gods and goddesses, the vegetation became a manifestation of the divine. The gloomy depths of the forests and groves elicited fear, awe, and reverence in the human minds. The human imagination conjured images of supernatural figures like dryads, nymphs, and fauns haunting the forests and glades. But more importantly, the humans believed the forests and groves to be the sacred place where the highest divine powers manifested in the form of the Mother Goddess.

The sacred character of trees and vegetation can be traced among the earliest races of the world. Comparative mythology of the Aryans, Greeks, Sumerians, Hebrews, and Africans mention of the created world's early days, when the earth and all natural elements seemed to be endowed with a mystical vitality. Accounts of sacred ruins containing sacred trees and groves have appeared across various cultures and religious faiths. Apart from being the objects of worship, trees have been intimately connected with temples and other sacred places of worship. Commenting on the sacredness of vegetation, Nanditha Krishna and M. Amirthalingam (2014, 18) explain:

> Sacred trees form an important part of the ecological heritage of India. Most temples, towns, and villages—and sometimes even Sikh temples and Muslims *dargahs*—are associated with trees. Some plants are sacred to the individual deity; others are sacred to the place. Sometimes, the tree is an integral or even larger part of the sanctity of the shrine; towns and cities and dynasties have been named after sacred trees.

The same could be said about the Greeks and other cultures from around the world. Many trees were revered as sacred in the complex polytheistic religious system practised in ancient Greece. Most of the Olympian divinities had sacred trees, animals, and birds associated with their worship. Apart from the Olympian divinities, lesser divinities like nymphs, satyrs, dryads, and other spirits were worshipped in sacred groves and forests.

Local cults also flourished in ancient Greece; they propitiated lesser-known divinities who had little to no connection with the Olympian deities. The roots of those cults predated the worship of the Olympian gods and goddesses. These cults emerged due to the ancient faith of honouring the earth.

In the Scandinavian culture, the world has been pictured as a gigantic tree. Similar motifs can also be found in other cultures from around the world. The "Tree of life" is a recurrent motif worldwide; gods, animals, and humans are often described taking birth from trees. "It is certain that, to the primitive religious mind, the tree (or rather, certain specific trees) represented a power" (Eliade 1958, 268). The earliest faiths never separated religion, nature, and the many symbols that the human mind associated with the natural world. Explaining the emergence and function of trees in religion and consciousness, Mircea Eliade writes that, "a tree impressed itself on the religious consciousness by its substance and its shape, but this substance and this shape became significant just because they impressed themselves on the religious consciousness, because they were "singled out", because, that is, they had "revealed" themselves" (Eliade 1958, 268). The veneration of trees, forests, and groves was largely due to the symbolic beliefs signified by vegetation life. Behind the symbols, there was always the presence of a spiritual being.

The Great Mother and Her Groves

The forests, groves, trees, and vegetation symbolized a primal creative force- the feminine principle responsible for the creation of all life forms. This feminine principle existed at the dawn of human civilization, even during the neolithic age. Women and feminine principles in particular were associated with the land's fertility and abundance of vegetation. "The fertility of the earth is bound up with feminine fecundity; hence women become responsible for the abundance of harvests, for they know the 'mystery' of creation. It is a religious mystery, for it governs the origin of life, the food supply, and death" (Eliade 1981, 40). Through numerous myths and cultural accounts, Mother Earth has been credited with bringing forth the creation by herself through parthenogenesis. In Greek mythology, "the first deity is female, a basic, matriarchically concept of mother earth and her fertility as primary and divine; comparative studies of iconography from primitive societies provide abundant evidence to confirm this archetype of the primacy of the feminine" (Morford and Lenardon 2003, 54). The earth and its fecundity were of particular interest to the primitive human communities. She was the bountiful bestower of life, and with the evolution of religious beliefs, cults honouring the earth have also undergone changes. But the core ideals honouring the earth and its fertility have remained unchanged. Shedding light on the origins and evolution of the worship of the Great Mother, Lotte Motz (1997, 6) writes, "since women participated more fully in tilling the soil than in hunting savage beasts and may even have invented agriculture, a matriarchal society may have arisen, in which it is only natural that the mother goddess, now seen in the image of the fruitful earth, the Earth Goddess, should have acquired yet another powerful dimension." The Great Mother goddess was the anthropomorphic embodiment of life, death, and regeneration. With the evolution and progress of human civilization, "the function and attributes of the mother goddess multiplied with the increasing complexity of social organization; still nurturing the wild creatures of the forest, she would now rule domesticated beasts and the seed placed in the soil" (Motz 1997, 6). In Minoan culture the Great Mother has been represented through various religious and artistic motifs, portraying her variously as the mother of beasts and vegetation.

In India, ecology, nature, and cults of the Mother Goddess have prevailed since prehistoric times. The hunter-gatherer tribes of India worshipped the Great Mother in various forms to ensure a bountiful harvest and abundance of prey. Like her Minoan counterpart, the Great Mother venerated by Indian communities was closely linked to the earth and nature. She was the mother of vegetation and animal life, and human communities sought her grace for their survival. During the Vedic age, her cults underwent a transition; she was no longer the primitive mother of all things. Rather she came to be worshipped in different forms. In this context, Nanditha Krishna (2017, 29) adds:

> One of the most beautiful hymns of the *Rig Veda* is dedicated to Aranyani, the goddess of the forest. She is an elusive spirit, fond of solitude, and fearless. The poet asks her to explain how she can wander so far from civilization without fear or loneliness. He creates a beautiful image of the village at sunset, with the sounds of the grasshopper and the cicada and the cowherd calling his cattle. She is a mysterious sprite, never seen, but her presence is felt by the tinkling of her anklets and her generosity in feeding both man and animal.

Aranyani was the goddess presiding over plants, vegetation, and wildlife who ensured that humans and non-humans co-existed harmoniously. She was worshipped alongside Prithvi the earth goddess, whose worship was extensively practised for the land's fertility. "She is the source of all plants, especially crops, and also nourishes all creatures that live upon her. She is described as patient and strong (12.1.29), supporting the wicked and the good, the demons and the gods" (Kinsley 1988, 9). In this regard, Prithvi embodied the very aspects the Greek Gaia represented. In other cultures of antiquity like the Egyptian and Babylonian, the Mother Goddess occupies a prominent place of worship in the religious and cultural system. For the Egyptians, Isis was the Great Mother who shared complex and interconnected ties with other gods like Osiris and Horus. In the words of Lesley Jackson (2016, 11), "the interconnectedness of all life is central to the concept of the Great Mother and Isis becomes increasingly connected through her relationships with the other deities and her assimilation of other goddesses." For the Nordic tribes, she was Freya while the Sumerians and Babylonians hailed her variously as Inanna and Ishtar.

A crucial aspect of these cults included the reverence of the ecology; the Great Mother was the patroness of all life forms, earth, and every conceivable thing in nature. From the neolithic tribes to the followers of mystery religions, the veneration of the Mother Goddess was an important aspect not just of the religious beliefs, but also of the cultural practices.

The earliest forms of nature worship typically involved the veneration of different objects and symbols found in the environment. This also led to the creation of sacred groves; a tree was marked for worship and the place was declared sacred and inviolable. According to Nanditha Krishna, "sacred groves are the home of local flora and fauna, a veritable gene pool of animal, insect, bird and plant species and a mini-biosphere reserve. The rich plant life helps to retain subsoil water and, during the hot summer months, the pond in the grove is often the only source of drinking water" (Krishna 2017, 38). In ancient Greece, the groves were sacred sites where altars dedicated to gods and goddesses were situated. In the pre-Hellenic Bronze age, the cult of the Great Mother predominated the worship of most other divinities. Her worship was frequently carried out in groves and sacred trees were venerated as her symbols and totems. For the Romans, she was Terra, the very embodiment of the earth. The Roman paganism was very much rooted in Greek faith. Greek gods and goddesses had their Roman counterparts who shared similar attributes. Among the innumerable rural and pastoral communities of Greece and Rome, which depended predominantly on farming and animal husbandry, the ancient beliefs and cults retained their significance. Among the local communities who remained beyond the influence of princely hegemony, the various cults remained active, worshipping trees and other natural symbols deemed sacred to the Great Mother. Noted historian Pausanias has documented the rituals and rites practiced by the people of Mysaeum honouring Demeter, an embodiment of the earth goddess; Pausanias (1910, 372) details as such:

> In the Mysaeum is a grove: trees of all kinds grow in it, and water wells up in plenty from springs. Here they keep a seven days' festival in honour of Demeter. On the third day of the festival the men withdraw from the sanctuary, but the women stay behind and perform by night the rites which custom prescribes. Not only the men, but even dogs of the male sex are turned out of the sanctuary. Next day the men come to the sanctuary, and the women laugh and jeer at them, and they at the women.

This ancient ritual illustrated the sacred importance of the female in the Greek culture and religious faith. Female divinities like Gaia, Hera, Demeter, and Artemis diversely embodied the different aspects of the natural world. And nowhere else was the ecology so closely related to religion as in the sacred groves. Associating an element of sacredness to the plants and groves reveals the social, economic, cultural, and religious concerns of the human communities. With the progress of civilization, agriculture began to gain new-found importance. Large tracts of forests were cleared for farming purposes. But the sacred groves were considered inviolable; no trees could be cut or wildlife hunted in the groves declared sacred. "Sacred groves are the *tapovana* that once existed within the forests of ancient India, where the ashramas of the rishis were located. The *tapovanas* were inviolate, unlike the *mahavana* where man encountered and hunted powerful predators, or the *shreevana*, which were exploited for their resources" (Krishna 2017, 38). For the Chaldeans, trees were the abodes of divine beings and it was a grave sacrilege to cut one down. "In a bilingual hymn, which is of Accadian origin, and probably one of the most ancient specimens of literature in existence, a mystical tree is described as the abode of the gods. And it was probably by a similar process of idealisation that a conventional representation of the sacred tree came to be one of the most important symbols of Chaldaean religion" (Philpot 1897, 4-5). The earliest shrines typically involved crude stone images placed under sacred trees in a grove. Offerings were made before the image to propitiate the divine being it represented. In Palestine, the Canaanites' altars were always present beneath sacred trees. Following the Canaanites, the Israelites also set up their sacred altars beneath a tree which was known as the *ashera.* This sacred tree represented the Mother Goddess who went by the name of Astarte or Ashtoreth. Subsequently, with the evolution of religion, the *ashera* became associated with other significant divinities like Jehovah as Astarte's cult went into decline.

The trees worshipped by the human community not only became a part of the religious system, but also made their way into the fabric of local mythology. It is worth mentioning that myths and religion have been inseparable over the ages. The justification of many religious beliefs and rituals can be traced in the community's mythology. A plethora of examples can be found in the annals of mythology. The laurel tree deemed sacred to Apollo in Greece is regarded to be the body of his beloved Daphne, who spurned his advances. This episode has been rendered in vivid detail by Michael Grant and John Hazel (2005, 163):

> It was through his mockery of Eros, the god of love, that Apollo fell in love with Daphne. For after he had unfavourably compared Eros' flimsy weapons and slight stature with his own prowess at archery, Eros, to punish him, loosed two shafts from his bow from the top of Parnassus. One pierced Apollo's heart with its gold tip and made him desperate for Daphne; the other, a blunt arrow with a leaden tip, struck Daphne and made her impervious to any lover. Apollo pursued Daphne through the woodlands until, on the banks of her father's river, the Peneius, he very nearly caught her. Here she prayed fervently to the river-God to save her, and was at once rooted to the spot and changed to a laurel tree, to which she gave her name. Apollo had to abandon his courtship but, as God of music and the bow, decreed that henceforward a laurel garland should decorate his lyre, his quiver, and the heads of minstrels.

Daphne was a nymph who got transformed into a laurel tree, implying the links between vegetation, women, and sacredness. The association of the laurel tree with Apollo came at a later stage in Greek history, when the older cults were replaced by the Olympian faith. A case to point would be the sacredness of the oak tree. In Greek religion and popular culture, it was considered sacred to Zeus. But originally, the oak was sacred to the goddess Dione. Despite the growing importance of Olympian divinities, the association between women and trees never lost its meaning for the Greeks. Along with Gaia, Rhea, Hera, and Artemis, Demeter is another female divinity intimately connected with the earth and vegetation. The Greeks propitiated her for bountiful harvests and groves sacred to her were inviolable spaces. A curious myth narrates the importance of trees and sacred groves in the worship of Demeter. Erysichthon the King of Thessaly was impudent enough to cut trees from Demeter's sacred grove. He needed wood for his banquet hall and paid no heed to warnings from by-standers. The anguished nymphs and tree spirits residing in the grove called upon Demeter who appeared before Erysichthon in the guise of a priestess. She implored him to cease his destructive act, but her pleas fell on Erysichthon's deaf ears. Consequently, Demeter cursed him to suffer from insatiable hunger. Erysichthon spent all his wealth to appease his hunger, and eventually devoured his own flesh and died.

In Hellenic faith, Demeter was the goddess of vegetation and fertility, assuming many of the attributes that had characterised Gaia and other mother goddesses. In Hellenic depictions, Demeter has most commonly been associated with grains and the corn in particular. But her worship with Persephone and the origins of the Eleusinian Mysteries points towards the more ancient vegetation myths, prevalent in different cultures. In this sense, Demeter shared most of the attributes of the Great Mother. When Demeter's daughter, Persephone was abducted by Hades, she went into mourning and the whole world suffered on account of her grief. Crops failed and vegetation withered away. Zeus implored her to restore her blessings on the world and save all life forms from famine, hunger, and death. Persephone was allowed to spend half of the year with Demeter, while the other half, she spent with Hades as the Queen of the Underworld. Demeter showered her blessings on the earth during the time Persephone spent with her. "The myth of Demeter and Persephone represents another variation of a fundamental and recurring theme—the death and rebirth of vegetation as a metaphor or allegory for spiritual resurrection" (Morford and Lenardon 2003, 318). The spiritual significance of forests, groves, and vegetation has been extensively discussed in Hindu culture, religion, and myths. The idea of forests and groves as sacred spaces has remained ingrained in Hindu beliefs. "The worship of plants is an ancient phenomenon in India. It is probably the oldest form of worship. The association of a single tree with a sacred *sthala* or *sthan* is reflected in the *chaitya vriksha* and *sthala vriksha* of literature and society" (Krishna 2014, 19). In India, there exists a strong relationship between ecology and socio-economic dynamics. Plants were revered for their medicinal and economic values; by ascribing them with spiritual and sacred importance, the human communities worked towards preserving the plants.

For the various communities and tribes of India, trees were the abodes of divine beings. The *Ashvattha vrikhsha* or the Pipal tree was the sacred abode of the various gods and goddesses who were believed to reside in the tree's leaves. In the *Rig Veda*, the *kalpa vriksha* finds frequent mention. It has been mentioned as the divine tree that emerged out of the ocean when the gods and demons churned it. The gods claimed it for themselves and the tree was consecrated in the gardens of heaven. A symbol of life, the tree supposedly has the powers to fulfil every kind of wish and bestow prosperity on the human community.

The *Mahabodhi* tree on the other hand, bestows wisdom and enlightenment. Associated with Lord Buddha, the Mahabodhi tree dispels the darkness of ignorance and leads the true seeker down the path of knowledge. The sacredness of trees in Indian religion and cultural beliefs testifies to the inherent importance of vegetation for the human community. On account of this, several trees have acquired the tag of *kalpa vriksha* in India. These include the banyan tree, coconut tree, and pipal tree. Explaining the significance of the *kalpa vriksha*, Nanditha Krishna (2017, 120) adds as thus:

> The *kalpa vriksha* focuses the mind in a spiritual direction. It gives and never disappoints, whether they are spiritual or materialistic demands. However, it never grants evil desires. It is a means to reach the gods. Unlike human beings who ask for and desire material objects, the *kalpa vriksha*, like all trees, gives food, shade, fruits, nut and timber, besides cleaning and purifying the air. It protects the traveller against heavy winds and stormy rains and is self-sufficient.

In India, it is unsurprising to find sacred groves near the villages, where the presiding goddess was consecrated. Through an amalgamation of myths, folklore, and legends, Indian communities inextricably tied together different aspects of religion, cultural beliefs, social practices, and environmental consciousness. "Sacred groves in India may belong to village communities, state forest and revenue departments, temples, tribal communities, gram panchayats, and even private individuals, as in Kerala. They are protected by local communities through taboos and sanctions that have cultural and ecological implications" (Krishna 2014, 82). For the indigenous communities, the forests and groves were the sacred abodes of the gods and goddesses. As the groves were the properties of the divine beings, its sanctity was of paramount importance and every effort was made by the human community towards its preservation and conservation. Explaining the close ties between the ecology and the Great Mother, David R. Kinsley (1988, 179) explains:

> The identification of the Devi with matter, the earth, or the cosmos is often expressed by identifying parts of the world with parts of her body. The *Devi-bhagavata-purana* calls the earth the Devi's loins. The same text speaks of the oceans as her bowels, the mountains as her bones, the rivers as her veins, and the trees as her body hair. The sun and moon are her eyes, and the nether worlds are said to be her hips, legs, and feet.

The sacredness of the earth and nature has been a fundamental aspect of Indian religious and cultural beliefs. Around this inherent belief in the sanctity of the earth and nature, several cults pertaining to the Mother

Goddess have flourished in India. Nearly in every Indian village, the Mother Goddess is worshipped as the tutelary deity. Typically, she is honoured as the *grama devi* or the patron goddess of the village. She is known by various names, "such as *Mata*, *Amba*, *Amma*, *Kali*, *Rati* and so on; sometimes dreaded, sometimes protective, warding off evil influences and imparting fertility by virtue of her life-giving energies (*sakti*) (James 1959, 113)." Her cults functioned through a mixture of reverence and fear. Venerated as the Great Mother who sustained all forms of life and bestowed bounty and riches on the human community, she was also feared as the one who could curse the humans with diseases, droughts, and sterility.

In her aspect as the *grama devi*, the Mother Goddess is usually enshrined in a sacred grove or forest on the outskirts of the village. The image is consecrated beneath a tree which assumes sacred importance. This particular tree is known as the *chaitya vriksha*. Explaining its importance for the Indian communities, Nanditha Krishna adds, "the term *chaitya* denotes a tree with dense foliage and fruits which provides shelter. The tree was the chaitya or protector for sacred images before temples were built, hence the name" (Krishna 2017, 120). Considered to be the abode of the deities, these *chaitya vrikshas* were protected by religious beliefs. With the worship of the tutelary deity and the *chaitya vriksha*, the entire place assumed sacred importance. Other trees in the vicinity also became a part of the sacredness, thereby leading to the formation of sacred groves.

In later ages, the trees and groves became associated with *yakshis* or supernatural female beings vested with divine powers. The Indian communities propitiated those divine women to ensure the land's fertility. Not only were they believed to bestow fertility on the land, but also were believed to bless barren women with children. It was also customary on the part of the village communities to propitiate the goddesses to ward off diseases and misfortunes. "The outbreak of cholera or smallpox is attributed to an angry Mother, as is whooping cough and the various symptoms of demoniacal possession such as epilepsy. hysteria, delirium, convulsions and fever" (James 1959, 116). The earth goddess was first and foremost a mother, who brought forth all life forms. Subsequently, in her many aspects and avatars, she was variously worshipped as the goddess of fertility, agriculture, and the patroness of animals. "The invisible threads binding the plants, animals and men in a given place to the soil from which they spring, which gave them birth and nourishes them, are woven together by that life which beats in the mother as in her creatures" (Eliade 1958, 255). In this regard, the Great Mother Goddess or the Earth Mother was worshipped in most natural surroundings, namely the forests and groves, where the power of nature held sway.

Conclusion

In every religious and cultural history, we come across the presence of sacred trees which in turn became a part of the community's iconography. But the sacredness of trees, plants, and vegetation is rooted in primal mysticism and metaphysical beliefs which link the plant life to the female creative force. The earliest forms of nature worship involved the veneration of plants, stones, rocks, rivers, and other natural symbols which signified the intricate mysteries of the creation. Amidst all these, the primitive human mind could not go past the cyclical nature of fertility and vegetation. Plants and trees emerged from the earth and over the course of time withered away and died. This phenomenon appeared immensely similar to the cycle of life and birth of human beings. The fertility of vegetation mirrored the fertility of human women who were responsible for procreating and nurturing human life. This gave rise to the cults of the mother goddess, who embodied the earth's fertility and nurturing capacity. "The earth was seen by all primal people as the source of nourishment, protection, power, and the mystery of cyclic recurrence. Perhaps the first human analogy made was between the earth and the female, who performed the same functions on an individual level" (Sjoo and Mor 1991, 70). The ceremonies, rituals and myths originating out of the sacred cults of the Great Mother frequently rested on the intimate relationship between the human communities and the non-human nature. Environment and the land were considered to be the embodiments of the Great Mother who nurtured all life forms. David Leeming and Christopher Fee (2016, 7) recount the origins of the worship of the Great Mother Goddess and its close association with the land and fertility:

> For time out of mind, an ancient goddess reigned supreme, and the fertile soil of the land beneath her feet bespoke her immeasurable fecundity. Long ages she ruled, and many were her worshippers. Then, 5,000 years ago or more, perhaps in the region where steppe meets sea in the great central land mass where Asia and Europe conjoin, tribes began to shift and migrate, and their movements were to transform this goddess forever. These peoples began to move south and east into the Indian subcontinent, down into what is now Iran, up to the Baltic, west along the European coast of the Mediterranean to the Atlantic, and even across the ocean to lands past the horizon.

The representation of the Great Mother as the nurturer of life as well as a destroyer can be understood in the context of the natural environment. The early communities who initiated the cults of the Great Mother understood the role of nature as a force that sustained life as well as destroyed it. The benevolent Great Mother bestowed life and fertility through crops and at the same time, inspired awe and terror among the human communities through diseases and natural cataclysms. The reverence towards the ecology and the Mother Goddess in particular was manifested through a complex belief system, where the human communities worshipped the goddess through a multitude of symbols. No other symbol held greater sway than plants and trees which signified the cycle of life, death, and regeneration. Consequently, the Great Mother came to be worshipped through trees deemed sacred and harming the vegetation was considered to be an unpardonable sacrilege. As religion and environmental awareness became inextricably tied to one another, sacred groves gained importance in social and cultural practices. Human communities understood and appreciated the importance of vegetation; efforts were made to preserve trees, plants, groves, and forests through the imposition of religious, social, and cultural taboos. Creating sacred groves dedicated to the Mother Goddess, not only ensured the protection and conservation of trees, plants, and other vegetation, but also created an ecological microcosm, a miniature biosphere that sustained a diverse variety of life forms. Along with the vegetation, the animals, birds, and insects became a part of a scared system which sanctioned their conservation. The solemnity and seclusion offered by the groves seemed to the human consciousness a place of wonder and spiritual significance set apart by the *magna mater* to worship the very force that procreated and sustained life on earth- the Great Mother Goddess.

References

Ann, Martha and Dorothy Myers Imel. 1993. *Goddesses in World Mythology.* California: ABC-CLIO.

Baindur, Meera. 2015. *Nature in Indian Philosophy and Cultural Traditions.* London: Springer.

Berkes, Fikret. 2012. *Sacred Ecology.* New York: Routledge.

Burkert, Walter. 1985. *Greek Religion: Classical and Archaic.* Malden: Blackwell Publishing.

Collins, Derek. 2008. *Magic in the Ancient Greek.* Malden: Blackwell Publishing.

Cusack, Carole M. 2011. *The Sacred Tree: Ancient and Medieval Manifestations.* Newcastle: Cambridge Scholars Publishing.

Dowden, Ken. 2000. *European Paganism: The Realities of Cult from Antiquity to the Middle Ages.* London: Routledge.

Dugan, Ellen. 2003. *Garden Witchery: Magick from the Ground Up.* Minnesota: Llewellyn Publications.

Eliade, Mircea. 1958. *Patterns in Comparative Religion.* Translated by Rosemary Sheed. New York: Sheed and Ward.

Eliade, Mircea. 1981. *A History of Religious Ideas: From the Stone Age to the Eleusinian Mysteries.* Chicago: The University of Chicago Press.

Fee, Christopher and David Leeming. 2016. *The Goddess: Myths of the Great Mother.* London: Reaktion Books.

Gottlieb, Roger S. 2006. *The Oxford Handbook of Religion and Ecology.* Oxford: Oxford University Press.

Grant, Michael and John Hazel. 2005. *Who's Who in Classical Mythology.* New York: Routledge.

Graves, Robert. 2018. *The Greek Myths.* London: Penguin.

Hard, Robin. 2004. *The Routledge Handbook of Greek Mythology.* London: Routledge.

Hart, John. 2017. *The Wiley Blackwell Companion to Religion and Ecology.* Oxford: John Wiley and Sons Ltd.

Hesiod. 2004. *Theogony, Works and Days, Shield.* Translated by Apostolos N. Athanassakis. Baltimore, The Johns Hopkins University Press.

James, E.O. 1959. *The Cult of the Mother Goddess: An Archaeological and Documentary Study.* London: Thames and Hudson.

Kinsley, David R. 1988. *Hindu Goddesses: Visions of the Divine Feminine in the Hindu Religious Tradition.* Berkeley: University of California Press.

Knapp, Stephen. 2014. *Hindu Gods and Goddesses: Understanding Hindu Divinities.* Mumbai: Jaico Publishing House.

Krishna, Nanditha and M. Amirthalingam. 2014. *Sacred Plants of India.* New Delhi: Penguin.

Krishna, Nanditha. 2017. *Hinduism and Nature.* New Delhi: Penguin.

Larson, Jennifer. 2007. *Ancient Greek Cults: A Guide.* New York: Routledge.

Larson, Jennifer. 2016. *Understanding Greek Religion: A Cognitive Approach.* Abingdon: Routledge.

Lash, John Lamb. 2006. *Not in His Image: Gnostic Vision, Sacred Ecology, and the Future of Belief.* Vermont: Chelsea Green Publishing.

Mackenzie, Donald A. 1913. *Indian Myth and Legend.* London: The Gresham Publishing Company.

Macleod, Sharon Paice. 2014. *The Divine Feminine in Ancient Europe: Goddesses, Sacred Women, and the Origins of Western Culture.* North Carolina: McFarland & Company, Inc., Publishers.
Marinatos, Nanno. 2000. *The Goddess and the Warrior: The Naked Goddess and Mistress of Animals in Early Greek Religion.* London: Routledge.
Morford, Mark P.O. and Robert J. Lenardon. 2003. *Classical Mythology.* New York: Oxford University Press.
Motz, Lotte. 1997. *The Faces of the Goddess.* New York: Oxford University Press.
Nelson, Lance E. 1998. *Purifying the Earthly Body of God: Religion and Ecology in Hindu India.* Albany: SUNY Press.
Neumann, Erich. 2015. *The Great Mother: An Analysis of the Archetype.* Princeton: Princeton University Press.
Northcott, Michael S. 2015. *Place, Ecology, and the Sacred: The Moral Geography of Sustainable Communities.* New York: Bloomsbury.
Ogden, Daniel. 2007. *A Companion to Greek Religion.* Malden: Blackwell Publishing.
Pausanias. 1898. *Description of Greece.* Translated by J.G. Frazer. London: Macmillan and Co., Limited.
Philpot, J.H. 1897. *The Sacred Tree or the Tree in Religion and Myth.* New York: Macmillan and Co., Ltd.
Pogacnik, Marko. 2001. *The Daughter of Gaia: Rebirth of the Divine Feminine.* Forres, UK: Findhorn Press.
Porteous, Alexander. 2002. *The Forest in Folklore and Mythology.* New York: Dover Publications Inc.
Richer, Jean. 1994. *Sacred Geography of the Ancient Greeks.* New York: SUNY Press.
Riddle, John M. 2010. *Goddesses, Elixirs, and Witches: Plants and Sexuality Throughout Human History.* New York: Palgrave Macmillan.
Roman, Luke and Monica Roman. 2010. *Encyclopaedia of Greek and Roman Mythology.* New York: Facts on File.
Ruether, Rosemary Radford. 2005. *Goddesses and the Divine Feminine: A Western Religious History.* Berkeley: University of California Press.
Sjoo, Monica and Barbara Mor. 1991. *The Great Cosmic Mother: Rediscovering the Religion of the Earth.* San Francisco: Harper.
Sophocles. 2009. *The Theban Plays.* Translated by Ruth Fainlight and Robert J. Littman. Baltimore: Johns Hopkins University Press.
Taft, Michael. 2014. *Greek Gods and Goddesses*. New York: Britannica Educational Publishing.

Znamenski, Andrei A. 2007. *The Beauty of the Primitive: Shamanism and the Western Imagination.* New York: Oxford University Press.

CHAPTER IV

THE TUNES FROM THE SACRED GROVES: AN ECOCRITICAL READING OF *HIJAN HIRAO*

ROSY YUMNAM

The main aim and concern of ecocriticism are to explore the artistic and creative power of literature and to relate it to the environment in the present times. This is mainly aimed to bring about an awareness of environmental concerns so as to develop an understanding of the current environmental problems which may eventually lead to a change in the way one treats and nurture nature. Ecocriticism is believed to be one of the crucial recent research trends in the field of cultural and literary studies. It originated in the United States and currently, it has spread through various departments across the entire globe. Ecocriticism is defined as "an eclectic, pluriform, and cross-disciplinary initiative that aims to explore the environmental dimensions of literature and other creative media in a spirit of environmental concern not limited to any one method or commitment" (Buell et al. 2011, 418).

William Rueckert first used the term "ecocriticism" in the essay "Literature and Ecology: An experiment in Ecocriticism" where he tried to incorporate the study of ecology and literature. Furthermore, Cheryll Glotfelty defined ecocriticism as "the study of the relationship between literature and physical environment" (Glotfelty 1996, xviii). She elaborates and asserts that "Ecocriticism takes as its subject the interconnections between nature and culture, specifically the cultural artifacts of language and literature. As a critical stance, it has one foot in literature and the other on land; as a theoretical discourse, it negotiates between the human and the nonhuman" (Glotfelty 1996, xix). According to Howarth, an ecocritic can be viewed as "a person who judges the merits and faults of writings that depict the effects of culture upon nature, with a view toward celebrating nature berating its despoilers, and reversing their harm through political action" (Howarth 1996, 69). Toeing in the line of Glotfelty, Richard Kerridge is of the view that "The ecocritic wants to track environmental ideas and representations wherever they appear, to see more clearly a debate which seems to be taking place, often part-concealed, in a great many cultural spaces. Most of all, ecocriticism seeks to evaluate texts and ideas in terms of their coherence and usefulness as responses to environmental crisis" (1998, 5). Relatedly, Buell et al. contends that literature portraying environmental degradation and harm may not be able to directly bring about a change in the society but "reflecting on works of imagination may prompt intensified concern about the consequences of such choices and possible alternatives to them" (Buell et al. 2011, 418). This may in the process help in creating environmental awareness and instil a sense of the close ties that exist between culture and environment. Consequently, it can assist in averting the current environmental crisis. Concurrently, culture and nature are interdependent. So, examining the differences and similarities between cultural evolution and natural evolution remains a pivotal context to understand cultural ecology. Furthermore, cultural ecology is an approach of ecocriticism which aims to unify various cultures of knowledge across diverse disciplines and sub-disciplines. It explores textual knowledge to contribute knowledge about culture and the environment. Generally, human beings are cultural beings dwelling in an environment. So, they are instinctively in tune with nature. Zapf (2010, 138) asserts that,

> "Literature and other forms of cultural imagination and cultural creativity are necessary in this view to continually restore the richness, diversity, and complexity of those inner landscapes of the mind, the imagination, the emotions, and interpersonal communication which make up the cultural ecosystems of modern humans, but are threatened by impoverishment by an increasingly over economized, standardized, and depersonalized contemporary world."

Therefore, the need to explore the texts and discourses which can endear upon continuing the close relationship between human beings and nature.

Objectives and Methodology

This paper seeks to examine the ecocritical elements in *Hijan Hirao*, a folk song of Manipur in North-East India. *Hijan Hirao* is a long narrative poem which is sung on the closing day of the *Lai Haraoba* festival of Manipur. The sacred grove (*umang*) occupies a pertinent place in the religious landscape of the Meitei community of Manipur. Revered deities called the *umang lais* (sylvan deities) are believed to reside in the sacred groves. In the *Lai Haraoba* festival, the *umang lais* are given due respect and homage through various songs, dances and rites and rituals. *Hijan Hirao* is one such song performed during the festival. Sacred groves are the expanse of forest housing rich biodiversity religiously protected by the local people on account of their cultural, religious and customary beliefs. They are guarded and protected by the local people as they are believed to be the home of their deities. As a result, sacred groves are considered to be the repository of a rich variety of flora and fauna (Devi et al. 2005, 1542).

The Meiteis believe in ancestral worship in the form of the revered *umang lais* in the forest areas. There are around 365 *umang lais* (Umang Lai) in Manipur. The *umang lais* are associated with a range of beliefs and taboos which comes across with the various religious practices. In a way, these diverse social restrictions in the concept of sacred grove aid a great deal in conserving the abundant flora and fauna of the forest. To please the *umang lais*, *Lai Haraoba* is performed every year to honour and gain their favour. During the festival, indigenous rights and rituals in the form of prayers, songs and dances are performed in these sacred groves. In a way, this helps in conserving the environment. Sacred groves spread over a wide ecosystem, are well preserved due to their religious and cultural significance. They are perceived to be a storehouse of a wide range of plants, trees and valuable medicinal plants. The association of nature with the rich ecosystem and the sacred groves serve as a haven for threatened species. In light of these observations, it can be contested that the tunes of *Hijan Hirao* with its ethical orientation towards nature can embark upon the ecocritical stance of preserving and conserving culture and the environment.

The study is analytical in the method. It tries to examine the ecocritical elements in *Hijan Hirao*, a folk song of Manipur. The framework of cultural ecology will be explored to read the folk song. As ecocritics dwell on the role of conserving ecology through texts, the paper aims to look at how the tunes of *Hijan Hirao* can inculcate the noble thought of conserving the environment and culture in the minds of the readers. The study is based on primary and secondary sources, research journals and online sources.

Literature Review

The main concern of ecocriticism is to create an awareness of the environmental problems currently experienced in the present-day world. Ecocriticism uses literature as a tool to endeavour the rising environmental crisis and aims to bring a change in the way one perceives and treats nature. Ecocritics are of the view that nature existed before the existence of human society or language. As Raha (2019) asserts, "Ecocriticism applies a range of ecocritical concepts, such as, growth and energy, balance and imbalance, symbiosis and mutuality, sustainability and unsustainability" (1). Hence, ecocriticism endeavours to establish a deep bond between humans and nature through various texts. In cultural ecology, human culture is considered to be mutually dependent on natural energy cycles and ecological processes. Zapf is of the view that, "Even as the dependency of culture on nature, and the ineradicable presence of nature in culture, gain ever more interdisciplinary attention, the difference between cultural evolution and natural evolution is increasingly acknowledged by cultural ecologists" (2010, 137). Further, Zapf in the book, *Literature as Cultural Ecology* asserts that "literature can itself be described as the symbolic medium of a particularly powerful form of "cultural ecology"" (2010, 138). Literary texts have been examined to study the feedback relationship of the current cultural systems with the needs of human and nonhuman nature (Zapf 2010, 138). This "paradoxical act of creative regression" (Zapf 2010, 138) made cultural renewal possible (Zapf 2010, 138). Literature in the context of cognitivity and creativity can be understood through "a cultural-critical metadiscourse, an imaginative counterdiscourse, and a reintegrative interdiscourse" (Zapf 2010, 138). The textual form delves into the fossilised "social structures and ideologies, symbolically empowers the marginalized, and reconnects what is culturally separated" (Zapf 2010, 138). This helps in maintaining a balance in the sphere of political, economic or practical forms of understanding human life in relation to the world around (Zapf 2010, 138). Therefore, literature can be perceived as cultural-ecological knowledge which collates realistic, pragmatic knowledge and also relates it to scientific accounts of ecology (Zapf 2010, 138). Culture and nature are closely associated and this relationship plays a pertinent role in the field of literary culture. The literary cultural discourse may be in the form of myths, legends, pastoral literature, rites and rituals, nature poetry, fairy tales and proverbs which have become an important medium to connect to one's culture, traditions and history. Therefore, the close interconnectedness between nature and culture, human and nonhuman forms has become a significant means to delve deeper into understanding the meaning of life pertaining to the surrounding environment.

Various researches have examined the significance of the use of ecocriticism as a literary tool to analyse diverse literary works. Research in the field of literature, culture and nature has been generally found to be examined from the perspective of a cultural-ecological approach. The textual, cultural and natural bond is an important element of ecocriticism. This means that nature can be examined, creatively used, interpreted, represented or empowered in various forms of culture through textual practice. Sankyan and Sigroha (2017,

47) explored the ecocritical elements in the writings of Temsula Ao and Kynpham Sing Nongkynrih, two poets from the North-East region. The study delves into the umpteen ecological glory as reflected through their poems and also shows the vital role played by literature to bring about environmental consciousness. The ecocritical elements of the Ramo community analysed by Koje and Mishra (2020, 43) show that their traditions and culture are influenced by the ecological environment. It also explored the deep bonding which exists between human beings, animals, nature and the spirits. The analysis of the ecocritical elements in Jayanta Mahaprata's selected poems brings forth his intimate relationship with nature which helps in creating an awareness of environmental conservation (Ranjit and Rajkumar 2020, 2). Relatedly, Fajar (2019) asserts the importance of local wisdom and folklore in its role to educate the next generation with the rising need for ecological education. Another study on the ecocritical elements in the folktales of Northeast India shows the close interconnectedness between human beings and flora and fauna as the main feature of tribal writing (Raha 2019, 3). It also explains another vital feature where natural phenomena are described and examined in an aesthetic manner. Therefore, the preceding observations support the significance of an ecocritical understanding of the diverse texts and their crucial role in creating environmental awareness among people. The close association of nature with human beings which induces in them a sense of belongingness is one of the binding features of the diverse texts discussed above. This will assist in bringing a sense of thoughtfulness and moral responsibility towards nature and the world around.

Hijan Hirao: An Ecocritical Analysis

Glottfelty (1996) observes "Describing a poem as stored energy, Rueckert explains that reading is an energy transfer and that critics and teachers act as a mediator between poetry and the biosphere, releasing the energy and information stored in poetry so that it may flow through the human community and be translated into social action" (xxviii). William Rueckert stresses on the vital role of poetry to induce environmental awareness among people in society. Poetry endears a deep bond between humans, tradition, culture and the world around them. Ecocritics are of the view that the traditional folklore, culture and history bridge the ecological truth with the people in the society. This aids in creating a certain connection with the past glories and helps in reconnecting with the diverse environmental concerns currently faced today. In this context, the tunes of *Hijan Hirao* serve as a means to nurture environmental concerns. The folk song, *Hijan Hirao* is sung on the concluding day of the *Lai Haraoba* festival of Manipur.

Lai Haraoba, according to Laisram (2009), "is a living tradition of the Meiteis" (169). Relatedly, *Lai Haraoba* being the religious and social festival of Manipur, is not only a storehouse of diverse aspects of the culture of Manipur but also serves as the repository of a variety of Manipuri songs (Singh 2003, 12). The folk song, *Hijan Hirao* is written in Meitei Mayek, transcribed and rendered into modern Manipuri and translated into English. It can be traced as early as the 13th or the 14th century CE (Singh 2015, 9) and the folk song is undated and anonymously written. The sacred song is in the form of a lament of the tree parents for their loving tree son. This folk song in the form of myth narration has been an inherent part of the *Lai Haraoba* festivals. The incorporation of the myth narration of *Hijan Hirao* in *Lai Haraoba* is to assimilate the close association of nature and culture in the context of the Meitei community. The ritualistic ceremony offered to a 'tree' before being cut down is the main theme of the poem. It is in the sacred groves that the *umang lais* reside. These sacred groves abound not only in religious significance but also showcase the innate relationship between humans and nature. The Meiteis believed that the disturbance of the sacred grove will draw wrath to the *umang lai.* So, cutting trees and forests is prohibited in and around the sacred groves. The Meiteis believed that nonhuman forms like trees and plants have souls. So, if one has to cut trees or plants for some purpose, one should do so by making some offerings to show reverence to the nonhuman forms.

The narrative of *Hijan Hirao* begins by describing the boat-making event of a Meitei king belonging to the Luwang clan, Luwang Ningthou Punsiba who is believed to have reigned around the 5th century to the 7th century CE. The sound of nature as evident through the gushing of the Singtha River and the images of insects made a remarkable introduction which brings forth its ecocritical connotation. It can be seen through the following lines:

> "Torential rain that poured amidst thunder and lightning
> Floods every lane, every meadow of the land.
> The Singtha River also is brimmed over
> And from it every crevice, every burrow
> Gushes down hiding snakes, grasshoppers, and crickets.
> Providence brought Luwang Ningthou Punsiba
> Towards the furious flow of the Singtha from his stroll
> And, Lo and behold! A grasshopper crosses the roaring river
> Clinging onto a tiny twig, half immersed in the water,
> A long-tailed rat performs the same feat
> Hanging onto a reed swept down by the Singtha."

(Trans. Singh 2015, 35)

The use of onomatopoeia and alliteration in the preceding lines heightens the rhythm and sets up the tone of the poem. In such a midst of nature which abounds in natural cadence, the King commends his two reliable artisans to make a boat. So, to fulfil the King's task, the members of the Luwang clan along with the two artisans set out to a hill called *Kouba Ingen.* The tradition of the time was to perform a certain ritualistic ceremony to revere the 'nonhuman tree' before cutting the tree. Thus, rituals rich with offerings, prayers and sacrifices were performed to the 'nonhuman tree' as a mark of respect for a living being. The use of pathetic fallacy in the poem intensifies human being's reverence of the ecosystem.

William Howarth's view on the root of ecocriticism can be detailed as follows, "Ecocriticism is a term derived from Greek words Oikos and Kritis, and in tandem, they mean "house judge." "Oikos" means "household," an interconnection of humans, nature and the spirit. "Kritis" means judge, "the arbiter of taste who wants the house kept in good order" (Glotfelty and Fromm 1996, 69). William Howarth's view on ecocriticism includes a profound concern and thought for nature. There arises a strong need for nature and nonhuman forms to exist in harmony in society. Nature and culture are interconnected and flow in the same direction. In the poem, the tree is considered to possess soul and consciousness. In narrating the story, the mother of the tree, Irok Laicham Leima bemoans the impending misfortune of her beloved son and grieves upon how tenderly she had nurtured him from a seedling to a full-grown tree. The mother tree's heart-wrenching cries for her loving son's misfortune can be compared with any other human being's grief-stricken sorrows at the thought of losing their beloved child. The mother tree bewails:

> "O my son, my precious,
> They have recompensed with gold and silver for your life
> They have made axe marks on your body.
> When the Sun wakes up to a new morning
> You will be hacked at your trunk, and
> The dismembered body of you, my son
> Will lie prostrate on the ground.
> I have traversed every slope, every glade and gorge
> But found none that matches you.
> O my son, my rarest treasure,
> Who can crush this abyss of despair within me?
> Who shall now replace you, the world in my eyes?"

(Trans. Singh 2015, 38-39)

The mother tree laments at the impending gloomy fate of her tree son. She is saddened that the king's men would cut down her mighty son the next day to make the *Hiyang Hiren* (the boat of a king). The uncontrollable anguish of the mother of the tree embodies the submissive voice recounting the passive pains, anguish and sufferings of the various plants, trees and many other nonhuman forms. The saddened emotions of the mother tree at the impending danger lying ahead for her beloved tree son is shown in the above stanza. The mother tree laments and tells her tree son that she had travelled every nook and corner of the world but she could not find someone as dear as he was. The phrase, "every slope, every glade and gorge" exaggerates the mother tree's laments of showing her deep sorrow laden emotions of her loving tree son. The phrase, "abyss of despair" exaggerates the depth of her sorrows. Hyperbole has been used in the above lines to describe the sadness of the mother tree for greater emphasis. She laments:

> "Shielded you from the cold icy gales from the North,
> Protected you from the ravaging rainstorms from above,
> Guarded you from the raging wild forest flames, and
> Raised you by me from a sapling.
> But all my hope has fallen apart
> Like wasted fruits and withered flowers.
> How do I endure this bottomless misery, this heartbreak?"

(Trans. Singh 2015, 39)

Furthermore, the father of the tree, Irok Ningthou soothes his son by wishing him to serve as the Luwang King's magnificent boat. The poem acts as a bridge to bond present-day human beings with that the ecological environment which is alienated in the world of today. The tree parents who are broken-hearted want their son to do his best to serve the humans even when he is cut down.

> "My dearest son, when you reach the land of the *Luwangs*
> Make sure that your trunk is placed as the head
> And your upper part as the toe of the deity.
> Your bare breast be encased with metals
> And your inside be waxed thoroughly.
> Let the boat you become be a *nungban*
> The boat which brings good health and prosperity.
> Let you become a race boat, a *hiyang* of all the *Luwangs*
> Which must be offered prayers and rowed every *Hiyangei*
> To herald new hopes and aspirations in the land."
> Filled with benevolence for the dearest son

> The grieving parents have their last look at him
> And return back as the sun dawns upon earth."
>
> (Trans. Singh 2015, 41)

The preceding stanza brings forth the beautiful rendering of human emotions towards the trees. At the impending cutting down of the tree son, the father tree consoles him and tells him to be brave. He told him to embrace all the necessities required to make him a boat of good luck and fortune. He further told him to become the majestic boat of the King of the Luwangs. The sorrowful parents after showering their blessings on their beloved son bade a heart-breaking farewell. The final glance at their loving son was filled with sadness and agony. Even today among the Meitei community, the boat of the king is revered and profound obeisance is offered to pray for good health and wealth of the land every year in the month of *Hiyangei* (month of November) in a traditional boat racing ceremony called *Hiyang Tannaba*. This reminiscence the cultural past of the Meiteis and brings forth the pertinent necessity of cultural communication with the present-day generation.

Furthermore, the cultural connection to the past can be shown through the invigorating myth narrations of the folk song. *Hijan Hirao* brings forth the beautiful cultural traditions of the Meiteis in its myth narration. The narration describes the regal boat of the king which is called *Hiyang Hiren* specially designed and carved for the purpose of the king's transportation and also for the boat race. In *Hiyang Hiren*, the flat part of the body, *hiru* is adorned with a *sangai's* head (*sangai* is a brow-antlered deer which is found only in the Keibul Lamjao National Park, Manipur, India). The tail of the boat, *hinao* is decorated with a person's head which is symbolic. The story of the embellishment of the beautiful *Hiyang Hiren* abounds in the mythical stories of the *sangai* and the figure of the symbolic head of a person which becomes a part of the Meitei history, cultural heritage, and literature. Till the present day, *hiyang tannaba* which is the Meitei boat racing tradition is celebrated every year in the month of *Hiyangei* (month of November) following the age long cultural traditions. The narrative captures the historical myth wherein the king, Luwang Ningthou Punsiba desired his loving brother, Luwang Pudongkhoi Khutkoiba to first board on the boat he wholeheartedly made. But his brother had already been transformed into a *sangai.*

"My brother who saw the sun and moon before me
Is my brother who has turned into an animal now
The antler of my beloved brother
Be placed at the hiru properly
So that he does not fall and get drowned."

(Trans. Singh, 2015, 28)

So, the king along with his nephew, Semba Mingamba, the son of his brother who had been transformed into a *sangai* went searching for his brother, the *sangai*. The son killed the *sangai*, as he was unaware of the identity of the *sangai.* Thus, full of remorse the king brought his brother, the slayed *sangai* by lovingly placing the body in the frontage of his boat. Even today, to venerate the mythical cultural traditions of the Meiteis, a wooden head depicting a *sangai* is used to decorate the front portion of the king's boat. Thus, *Hijan Hirao* sung in *Lai Haraoba* festivals kept alive the mythical cultural traditions of the Meitei community by singing the legend of *sangai*. The other part of the narration where a symbolic human head is placed at the tail of the king's boat also has its own mythical and historical recounting. In this narration, the historical account of the Meitei king, Thawanthaba (1196 CE - 1231 CE) contributes to the functional elements of the story. In a battle between the Meitei and the Khuman community, the Meiteis won the battle and so as per the custom prevailing at the time; the Meiteis took home the decapitated head of one of their leaders as a winning prize. Even today, to revere these heroic deeds of the Meitei king, a symbolic head depicting that of a person is kept at the tail of the boat of a king. The historical narrative in *Hijan Hirao* which is sung in *Lai Haraoba* serves as a way of recalling and recounting the remote past events in the cultural context of the Meiteis. As asserted by Ong (2004), aphorisms, proverbs, religious rituals and philosophical theory have all been strung and captured together in the memory of the humans manifested through their narrations. This alludes to the functions of narrative fostering its cultural and human existence. *Hijan Hirao* which is sung on the last day of *Lai Haraoba* festival with its mythical and historical narratives accounts for the significance of the cultural communication it conveys to society. In sum, the tunes of *Hijan Hirao* with its cultural and mythological implications can perceive to be the cultural-ecological knowledge which commends to collate the past, real and endeavours a practical ecological realm. The folk song, innately weaves together culture and nature through its mythical narrations. This distinctive feature assists in foregrounding the role of the folk song to inculcate much-needed ecological awareness among people. It has also contributed to delving deeper into comprehending the meaning of life in relation to the surrounding environment.

According to Zapf, "cultural ecology looks at the interaction and living interrelationship between culture and nature, without reducing one to the other. Literature is seen as a cultural form in which this living interrelationship explored in specifically productive ways, providing a site of critical self-reflection of modern civilization as well as a source of creative cultural self-renewal" (Zapf 2016, 1). One of the important characteristic features of ecocriticism is that it examines the interconnectedness between human culture and nature. It also explores and looks at how humans stay in tune with nature. More importantly, ecocritics espouse the role of nature in texts concerning human culture and nature and how these texts play a pivotal role in conserving ecology. The poem profoundly exemplifies the inherent bond which exists between nature, humans and spirits. The tree parents and their tree son are personified with human emotions. The poem narrates human beings' ethical orientations and values with nature in its varied forms like trees, streams, rivers, forests and hills. The poem clearly shows the harmonious relationship of the human form with nature. The offering of prayers and delicacies to the nonhuman tree before the tree is cut down shows the deep interconnectedness between the human and the nonhuman form. The trees and the plants in the sacred groves are religiously guarded and protected. They are prohibited from being disturbed and cut. However, under circumstances like the one in this folk song, the tree was required to be cut down as the King was in need of a boat for transportation. During this kind of situation, the King and his people performed sacred rites and rituals by offering fruits, delicacies and prayer to the revered tree. This clearly shows the ecocritical stance visible throughout the poem. The close connectedness between the King and his people and the nonhuman tree can be felt through the emotions exuded in the course of the poem. According to the religious beliefs and customs of the Meiteis, the sacred groves remain untouched by humans. During the *Lai Haraoba* festivals, the *umang lais* are worshipped through an array of songs and dances. The folk song, *Hijan Hirao* is also performed during the *Lai Haraoba* festival as a mark of obeisance to the *umang lai*. The performance of the song till the present day showcases the cultural communication it conveys to society. Through cultural communication and its ecocritical stance, the poem clearly stakes its claim of respecting, preserving and conserving the environment. Therefore, the folk song *Hijan Hirao*, reminiscing the past glories of the times when humans, the forest and the trees remained close and connected, can instil a noble thought of conserving the environment among people.

Conclusion

Ecocriticism examines the interconnectedness between human culture and nature. Ecocritics espouse the role of nature in texts concerning human culture and nature and how these texts play a critical role in conserving ecology. The tunes of *Hijan Hirao* through its ethical orientation towards nature build a formidable task of preserving and conserving the environment. The lament of the tree parents upon the impending cutting of their tree son is filled with human emotions. In the folk song, the vesting of human emotions through the nonhuman form establishes the close interconnectedness of human beings with nature. Concurrently, the ritual of offering prayers, flowers and fruits to the nonhuman tree gives reverence to nature. This forms a close cultural bond with nature and assists in creating a cultural construct of human civilization. Furthermore, as culture represents the ways of life of people through a range of beliefs and social practices, the cultural myth narration with a blend of historical accounts affirmed its cultural implications discernible throughout the poem. The literary cultural discourse in the form of myths and rites and rituals through folk songs has become a pertinent medium to connect to one's culture, traditions and history. The cultural traditions are reflected in its myth narrations where the boat of the king called *Hiyang Hiren* is described. The myth surrounding the *sangai* (a brow-antlered deer) and the human head which adorns the head and tail of the king's boat, *Hiyang Hiren* is a source of the Meitei history, literature and culture. To the present day, the Meitei boat racing tradition called *hiyang tannaba* is celebrated every year. Unquestionably, the tunes of *Hijan Hirao* which is sung on the concluding day of the *Lai Haraoba* festivals with its mythical, historical and cultural narratives bring forth the importance of the cultural communications and traditions it conveys to the society. As human culture is considered to be mutually dependent on natural energy cycles and ecological processes, the implication of culture through the myth narratives of *Hijan Hirao* endears to bring about ecological awareness among the Meitei community. The folk song clearly shows the harmonious relationship of the human form with nature. The love and the empathy showered on nonhuman form clearly stake its claim of an eco-critical stance. This close interconnectedness between nature and culture, human and nonhuman forms commend a deeper understanding into the meaning of life in the surrounding environment. As ecocritics dwell on the role of conserving ecology through texts, the cultural-ecological knowledge as perceived through *Hijan Hirao* can ignite the minds of the reader with the noble thought of sustaining and conserving culture and the environment.

References

Alexander, A. 2013. "Environmental Otherness: Nature on Human Terms in the Garden." *Otherness: Essays and Studies* 4(1): 1-28.

Armbruster, K and Kathleen R. Wallace. Eds. 2001. *Beyond Nature Writing: Expanding the Boundaries of Ecocriticism*. Charlottesville and London: University Press of Virginia.

Buell, Lawrence, Ursula K. Heise, and Karen Thornber. "Literature and Environment." *Annual Rev. Environ. Resource* 36, 2011: 417-440. Accessed January 26, 2022.
DOI: 10.1146/annurev-environ-111109- 144855.

Buell, Lawrence. 1995. *The Environmental Imagination: Thoreau, Nature Writing, and the Formation of American Culture*. Cambridge MA and London: Belknap Press of Harvard University Press.

Chandra, N.D. and N. Das. 2007. *Ecology, Myth, and Mystery: Contemporary Poetry in English from Northeast India*. New Delhi: Sarup and Sons.

Chhibber, B. 2008. "Indian Cultural Heritage and Environmental Conservation through Traditional Knowledge." *Mainstream Weekly* XL VI (25). Accessed January 24, 2022.
www.mainstreamweekly.net/ article746.html.

Devi, A, M.L. Khan, and R.S. Tripathi.2005. "Sacred Groves of Manipur, Northeast India: Biodiversity Value, Status and Strategies for their Conservation." *Springer*, 1541-1582, Accessed February 15, 2022. DOI 10.1007/s10531-004-0530-5.

Dreese, D. 2002. *Ecocriticism: Creating Self and Place in Environmental and American Indian Literatures*. New York: Peter Lang Publishing.

Fajar, D.A. 2017. "Eco-Literacy through Ecocriticism in Pekalongan Folklore." *Faculty of Languages and Arts Yogyakarta State University*, 73.

Fajar, D.A. 2019. "An Ecocritical Reading: Demonizing Nature Found in Batang's Folktale." Accessed January 27, 2022.
http://dx.doi.org/10.4108/eai.27-4-2019.2285346.

Glotfelty, C. 1996. "Introduction." In: *The Ecocriticism Reader: Landmarks in Literary Ecology*. Edited by C. Glotfelty and H. Fromm. London: University of Georgia Press.

Glotfelty, Cheryll and Harold Fromm. Eds. 1996. *The Ecocriticism Reader: Landmarks in Literary Ecology*, 105-111. London: University of Georgia Press.

Howarth, W. 1996. "Some Principles of Ecocriticism." In: *The Ecocriticism Reader: Landmarks in Literary Ecology*. Edited by Cheryll Gloffelty

and Harold Fromm, 69-91. London: The University of Georgia Press.
Kerridge, R. 1998. “Small Rooms and the Ecosystem: Environmentalism and DeLillo’s White Noise.” In: *Writing the Environment*. Edited by R. Kerridge and N. Sammells. London: Zed Books.
Koje, N. and K.C. Mishra. 2020. “Ramo Folksongs: An Ecocritical Study.” Accessed February 16, 2022. http://www.researchguru.net/volume/Volume%2014/Issue%201/RG9.pdf.
Laisram, Rena. 2009. *Early Meitei History: Religion, Society and the Manipur Puyas*. New Delhi: Akansha Publishing House.
Manes, C. 1995. “Nature and Silence.” In: *Postmodern Environmental Ethics*. Edited by Max Oelschlaeger, 43-56. Albany: State University of New York Press.
O’Brien, Susie. 2007. “Reading Ecocriticism in a Postcolonial Context.” In: *Five Emus to the King of Siam: Environment and Empire*. Edited by Helen Tiffin. New York: Rodopi Press. 177-200.
Ong, J. W. 2004. *Orality and Literacy*. London and New York: Routledge.
Raglon, R. and M. Scholtmejer. 2001. “Heading off the Trail: Language, Literature, and Nature’s Resistance to Narrative.” In: *Beyond Nature Writing: Expanding the Boundaries of Ecocriticism.* Edited by K. Armbruster and K.R. Wallace. Charlottesville: The University Press of Virginia. 248-262.
Raha, D. 2019. “Folk Tales of North-East India: A Tryst with Nature.” Accessed February 18, 2022. https://www.noveltyjournals.com/upload/paper/Folk%20Tales%20of%20North-2077.pdf.
Ranjit, G. and K. Rajkumar. 2020. “Ecocritical Elements in the Selected Poems of Jayanta Mahapatra.” Accessed February 15, 2022. https://doi.org/10.24113/ijellh.v8i5.10576.
Ricoeur, P. 1984. *Time and Narrative.* Vol. I. Chicago: The University of Chicago Press.
Rowland, S. 2012. *The Ecocritical Psyche*. London: Routledge Publication.
Sankyan, N. and S. Sigroha. 2017. “The Environment is Us: Humanities and the Ecological Crisis, An Ecocritical Reading of Poetry from India’s Northeast.” *Ikhtila - Journal of Critical Humanities and Social Studies*, 47-58.
Singh, Ch. Manihar. 2004. *A History of Manipuri Literature*. New Delhi: Sahitya Akademi.
Singh, P. K. 2015. *Hijan Hirao: Text, Context and Translation*. Kakching: Sahitya Seva Samiti.
“Umang Lai.” Accessed January 25, 2022.

https://artsandculture.google.com/entity/umanglai/g11fqpk235s?hl=en.

Westling, L. 2014. *Literature and the Environment*. Cambridge, Cambridge Publication.

Zapf, H. 2016. *Literature as Cultural Ecology*. UK: Bloomsbury Publication.

Zapf, H. 2002. *Literatur als kulturelle Ökologie: Zur kulturellen Funktion imaginativer Texte an Beispielen des amerikanischen Romans.* (Literature as Cultural Ecology: On the Cultural Function of Imaginative Texts, with Examples from American Literature.) Tübingen: Niemeyer.

PART-II

SACRED ECOLOGY AND SUSTAINABLE DEVELOPMENT

CHAPTER V

BIODIVERSITY CONSERVATION AND ECOSYSTEM SERVICES PROVISION OF SACRED GROVES IN AFRICA

JONATHAN C. ONYEKWELU

The practice of sacred grove system by indigenous people is an ancient tradition, especially in Africa and Asia. The historical development of sacred groves has been traced to pre-agricultural, hunting and gathering stage of human civilization, when human society was in its primitive state (Gadgil and Vartak 1975). All over the world, indigenous people have rich traditions of nature conservation through socio-religious and socio-cultural constraints on profligate use of common natural resources (Onyekwelu 2021a). Prior to the development of conventional natural resources conservation methods, various traditional methods have been practiced for centuries in Africa and Asia. Indigenous natural resources management practices evolved through the interactions between communities and their environment, thereby giving rise to practices and cultural landscapes such as sacred forests, sacred corridors and a variety of ethno-forestry practices.

Sacred groves are found in many countries across the globe, but predominantly in countries in Asia and Africa (Bhagwat and Rutte 2006; Dudley et al. 2010; Aniah and Yelfaanibe 2016; Sharma and Kumar 2021). They feature in various cultures (Africa, Asia, Europe, America and Australia) and occur in various forms such as sites for religious and cultural festivals, remnants of old forests, burial grounds for chiefs/kings, sites of ancestral worship, etc. (Onyekwelu 2021b). They have been noted to vary in size, from as low as 0.001 ha to thousands of hectares (as high as 26,326 ha) (Mgumia and Oba 2003, Khan et al. 2008). Sacred groves are viewed as residences of local deities and sites for religious/cultural rituals; thus, they are held sacred by indigenous people (Onyekwelu 2021b; Sen 2018). They act as a connecting link between humans, nature, socio-cultural heritage, religious and ethnic beliefs (Khan et al. 2008; Sharma and Kumar 2021). Today, sacred groves have become an integral part of tribal and rural lives and are revered for their religious, cultural, social, medicinal, economic, psychological and educational importance (Jhansi et al. 2018).

Before the advent of conventional natural resources conservation methods in the 19th century, many communities in Africa and Asia developed and practiced traditional methods of protecting both plant and animal species in their habitats. In Africa, indigenous people have rich traditions of nature conservation through socio-cultural and socio-religious use and respect for forests and some other natural resources like rocks, rivers, waterfalls, sacred corridors, etc. Up to 3,000 sacred groves have been identified in Ghana (Boadi et al. 2017). Gomé (2003 cited in Mouroufie et al. 2019) listed more than 6,702 sacred forests covering about 37,000 hectares in Côte d'Ivoire while 2,940 sacred forests, covering an area of 18,360 ha (or 0.2% of total land area of Benin) were listed in Benin Republic (Soury 2007). In Tanzania, a total of 920 traditionally protected forests were recorded in the Handeni district and Mwanga (Dudley et al. 2010). Borokini (2016) noted that almost all south-eastern and south-western communities in Nigeria have at least one sacred natural site while Adeyanju (2020) observed that sacred groves are symbols of identity for most Yoruba people in south-western Nigeria. There is hardly any community in Cross River State, Nigeria without a sacred grove, sacred pond or some other part of the environment that is set aside for the worship of the gods. Sacred groves have been reported in other African countries, including Burkina Faso, Mali, Ethiopia, Kenya, Tanzania, Togo, Morocco, etc. (Mgumia and Oba 2003; Aerts et al. 2006; Ajbilou et al., 2006; Kidegheso 2008; Kokou et al. 2008).

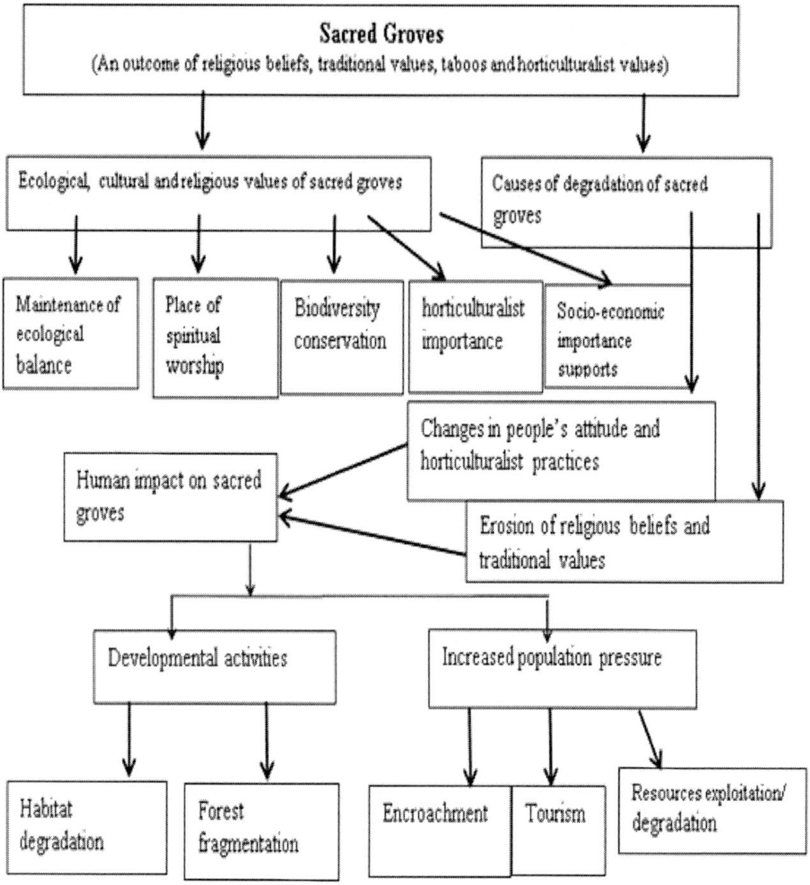

Figure 5.1: Representation of Sacred Groves: Ecological, Cultural and Religious Values and Relationship with Changing Traditional Beliefs, Human Impact and Causes of Degradation
Source: Adapted from Khan, M.L., D.V. Khumbongmayum, and R.S. Tripathi. 2008. "The Sacred Groves and their Significance in Conserving Biodiversity - An Overview." *International Journal of Ecology and Environmental Sciences*, 34: 277–291.

The success of sacred groves in delivering ecosystems goods and services to indigenous communities is hinged on their multifaceted dimensions and importance. Scientists are unanimous in their agreement that sacred groves serve social, religious, cultural, economic, ecological functions (Arora,

2006; Khan et al., 2008; Rutte, 2011; Onyekwelu and Olusola, 2014; Reuben and Kquofi, 2015; Sen 2018; Irvine et al. 2019; Parthasarathy and Babu 2019; Onyekwelu 2021b). Consequently, they have been able to deliver important ecosystem goods and services to the people. However, some of these important fulcrums of sacred groves are not only currently threatened, there are evidences that they are being eroded (Figure 5.1; Dudley et al., 2010; Ormsby and Bhagwat, 2010; Aniah and Yelfaanibe, 2016; Adeyanju 2020; Onyekwelu 2021a & b). Given these threats, the extent to which sacred groves would continue to provide their important functions now and in the future is uncertain. Onyekwelu (2021b) opined that more sacred groves may lose their sacredness as the threats confronting them escalate; as the interest of young people in sacred traditions reduces; as belief systems changes thus reducing the fear/respect for the gods; etc. Consequently, this chapter examines the importance (cultural, economic, ecosystem services and biodiversity conservation) of sacred groves, which has become necessary given the threats currently confronting them, especially in Africa.

Religious and Cultural Value

Religious and spiritual beliefs often contribute to social norms, acting as a potentially important instrument for environmental protection, which may be achieved through promotion of positive religious or spiritual associations with nature protection that motivates environmental consciousness (Reuben and Kquofi 2015). Almost all sacred groves in communities in Africa were established primarily to meet religious and cultural needs (Rutte 2011; Onyekwelu 2021a). This probably explains why most sacred groves are named after or dedicated to deity(ies), believed to be residence of deity(ies) and sites for religious/cultural rituals (Sen 2018; Onyekwelu 2021b). Sacred groves accommodate secrecy and privacy, which characterize the worship of deities in many communities (Probst 2013). In some indigenous communities in Ghana, the people believe that sacred forests are homes for their gods and the spirits of their cherished ancestors who need to be protected and worshiped so that the spirits will not be angry with them (Reuben and Kquofi 2015). A review of over a hundred studies in Africa and Asia by Dudley et al. (2010) led to the conclusion that sacred natural sites have spiritual values to one or more faith groups within their host communities. Thus, sacred groves are conserved for worship of traditional deities, practice of spiritual beliefs and religious rituals, cultural tenets and festivals of importance for sustenance of the cultural heritage of indigenous people.

Religio-cultural benefits are usually the main motivations for preservation of sacred groves. Onyekwelu (2021a) opined that deep religious reverence for nature, rather than resource scarcity, is the basis for commitment of indigenous people towards the preservation of sacred groves while Parthasarathy and Babu (2019) maintained that social, religious and environmental taboos enhance the sanctity of sacred groves and promote species conservation. Arora (2006) confirmed that socio-religious fencing greatly contributed to continued conservation of sacred groves. Chukwu et al. (2019) listed cultural beliefs, myths and traditional taboos among the measures that have been successfully used to preserve some sacred groves. The importance of sacred groves in community life is usually felt through religious-cultural practices (Onyekwelu 2021a). Most sacred groves have shrines on which sacrifices are made by priests, they are known to receive many visitors who seek supernatural interventions for their predicaments, such as healing from strange sicknesses, freedom from demonic oppression, overcoming fertility problems, poverty, joblessness, marriages, etc. (Daniel et al., 2016; Yusuf, 2016; Adeyanju 2020).

Reason for conservation of sacred groves			Traditional beliefs contributing conservation of sacred grove			Factors promoting biodiversity conservation in sacred grove			Socio-economics benefits derive from sacred groves		
Factors	Freq	%	Belief	Freq	%	Factors	Freq	%	Benefits	Freq	%
Tourism	6	10	Place of important community festivals	9	15	Respect for law and culture	3	5	Employment	6	10
Home of Deity	12	20	Belief in deity and/or taboo	24	40	Abomination to fell trees	12	20	Social amenities	–	–
Preservation of culture	9	15	Site for healing & the miraculous	12	20	Fear of deity, taboo/ superstition	12	20	Promotion of art & Culture	9	15
Nature preservation	3	5	Home of goddess & place of worship	15	25	Preservation of home of deity	30	50	Protection & Peace	9	15
Education & research	6	10	No idea	–	–	No response	3		Healing	18	30
Place of Worship	24	40							Revenue generation	12	20
									Tourism	6	10
Total	**60**	**100**		**60**	**100**		**60**	**100**		**60**	**100**

Table 1: Factors Contributing to Conservation of Sacred Groves and Ecosystem Services Provision from Osun-Osogbo Sacred Grove

Source: Onyekwelu, J.C. (2021a). "Significance of Traditional Forest Management System to Biodiversity Conservation and Ecosystem Services Provision: Case Study of Osun-Osogbo Sacred Grove, Nigeria." In: Felbermeier, Stimm and Seidl. eds. *Waldbau weltweit 2.0. Festschrift anlässlich der Verabschiedung von Prof. Dr. Dr. Reinhard Mosandl. Forstliche Forschungsberichte* 219: 132- 148. Munich, Germany.

Using a study in Osun-Osogbo sacred grove in south-western Nigeria, Onyekwelu (2021a) demonstrated the cultural and religious functions of sacred groves. For example, Osun-Osogbo grove was preserved because majority of the people (60%) in neighbouring communities see it as the home of the Osun goddess and a place of worship while 40% of the people pointed out that belief in deity and/or taboo was the primary traditional belief contributing to the conservation of the sacred grove (Table 1). According to 50% of the community dwellers, the preservation of Osun-Osogbo sacred grove as the home of the deity is the primary factor promoting biodiversity conservation. In a study on three sacred groves in south-western Nigeria, Adeyanju (2020) observed that most of the people in communities neighbouring Idanre Hills and Osun-Osogbo sacred groves strongly agreed that they are places of worship, protection and peace by the deity, provision of shelter for the deities, ritual sites, a place for promotion of indigenous arts and culture, among others (Figure 5.2).

Sacred groves are sites for rituals, initiations, festivals, and ceremonies, including the installation of traditional rulers (Daniel et al. 2016). In Nigeria, the Osun-Osogbo festival, domiciled at Osun-Osogbo sacred grove, has become one of Nigeria's biggest tourist attractions, with over 130,000 annual visitors, tourists, and pilgrims (Probst 2013; Onyekwelu and Olusola 2014; Adeyanju 2020). The Orosun, Ogun and Ije festivals at Idanre Hills sacred grove in Nigeria provide occasions for Idanre people to reconcile with their natural environment as well as recount historical experiences in local Idanre history and its broader Yoruba ideology, mythology and confederacy (NCMM 2007). During the Orosun festival, the priests and adherents embark on a trip to the sacred forest on Idanre Hills (Adebayo 2016 cited in Adeyanju 2020). In Igbo land of south-eastern Nigeria, sacred natural sites are regarded as the earthly abode of guardian spirits, where mystical and miraculous events could take place (Emeasoba and Ogbuefi 2013). Nigerian politicians are reported to visit these shrines for protection, mystic power, good fortune and long life (Emeasoba and Ogbuefi 2013).

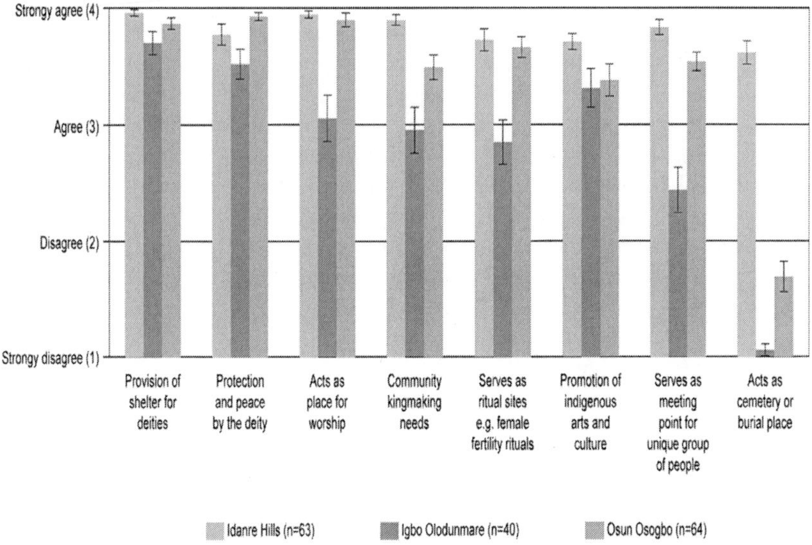

Figure 5.2: Religio-Cultural Benefits Driving Biodiversity Conservation in Three Sacred Groves in South-Western Nigeria
Source: Adeyanju, S.O. 2020. "Drivers of Biodiversity Conservation in Sacred Groves: A Comparative Study of Three Sacred Groves in South-west Nigeria." M.Sc. Thesis, Canada: University of British Columbia. 98 pages. Accessed July 6, 2022. http://dx.doi.org/10.14288/1.0389686.

In the Boabeng-Fiema monkey sanctuaries in Ghana, the Mona monkeys (*Cercopithecus Mona Mona*) are associated with local gods and are therefore considered sacred, thus it is taboo to kill them. Each sanctuary has a sacred area, where ceremonies and rituals are performed. Both Boabeng and Fiema have separate annual yam festivals that also celebrate the monkeys (Ormsby 2012). Due to effectiveness of the conservation of the monkeys, tourism population increased dramatically from 150 tourists in 1991 to 10,000 in 2005 (Ormsby 2012). Also, a taboo prohibits the hunting of the ursine black and white colobus (*Colobus vellerosus*) and the Campbell's monkey (*Cercopithecus campbelli lowei*). This taboo is underlain by religious beliefs as the primate species are considered by the indigenous people to be the children of the gods (Saj et al. 2006; Reuben and Kquofi 2015). The sacred groves in the *miombo* woodlands in Tanzania are reported to be central to the spiritual needs of the indigenous people, with specific trees and blocks conserved for cultural and ritual purposes (Morris 1995 cited in Mgumia and Oba 2003). The sacred groves are associated with ancestral and spirit worship; the burial sites of chiefs are particularly preserved and venerated (Shepherd 1992; Mgumia and Oba 2003). The Masai people of Kenya use forestlands as sacred sites for conducting prayers while the Gabbra pastoralists of the northern Kenya/Ethiopian borderlands are reported to have preserved forests as ritual sites (Schlee 1992; Mgumia and Oba 2003). Sacred forests are integral parts of their spirituality in the Ameru culture in Kenya, an abode for their ancestors who watches over them, which explains why they consider nature (especially sacred forest) as a sacred place for prayers, reflections, contemplation, health and wholeness (Mukaria 2021). Spiritual value was listed as one of the important functions of Church forests (sacred groves) in northern and central Ethiopia (Cardelús et al. 2012). The church forests, which are extensions of the Ethiopian Orthodox Tewahedo Church, are religious places as well as social centers that advance the material, spiritual well-being and social capital of the community by allowing social networks to develop and thrive (Orlowska and Klepeis 2018). The community dwellers are bound to their natural environment by the religious practices of the Ethiopian Orthodox Church.

Biodiversity Conservation Values

Biodiversity conservation is by far the most popular and studied function of sacred grove. By implication of their definitions, sacred groves are remnants of virgin (primary) forests that harbours rich biodiversity. Although they were not originally created for biodiversity conservation, sacred groves have contributed largely to in-situ biodiversity conservation, at both overstory and understory layers (Myers et al. 2000; Bhagwat and Rutte 2006; Khan et al. 2008; Kokou et al. 2008; Dudley et al. 2010; Ormsby and Bhagwat 2010; Onyekwelu 2021b; Onyekwelu et al. 2021; Zannini et al. 2021). Sacred groves are often rich in species, sometimes being similar or more species diverse than protected areas or forest reserves. Their remnant nature and high level of protection means that the areas protected as sacred grove often have particularly rich biodiversity. Usually, forest or species protection in sacred groves is achieved using taboos, religious beliefs or by dedicating the forest/species to deity(ies) (Onyekwelu and Olusola 2014). The sacredness attached to sacred groves by indigenous people and the fear of the deity(ies), which prevents or restricts people from felling trees and hunting animals, are largely responsible for the biodiversity conservation success of sacred groves (Onyekwelu 2021b). In addition, the pivotal traditional rules and taboos in sacred groves have preserved their integrity and served essential biodiversity conservation function (Parthasarathy and Babu 2019). Any sacred grove that loses its sacredness and the gods no longer feared will soon lose its integrity and may no longer be a sanctuary for biodiversity conservation. Today, sacred groves are noted to be reservoir of native biodiversity, refuge for endemic and endangered species and to represent biodiversity hotspots (Laird 1999, Myers et al. 2000; Kokou et al. 2008, Onyekwelu and Olusola 2014). According to Laird (1999), the importance of sacred groves as tool for biodiversity conservation is hinged on their widespread distribution and their important roles as reservoirs of local biodiversity of threatened species. Thus, the ecological importance of sacred groves as complementary biodiversity conservation method with associated environmental benefits is today being widely recognized (Salick et al. 2007; Ballullaya et al. 2019).

Forest type		Location	State of forest	Biodiversity Indices					Tree Growth variables			Author(s)
				NoF	NoS	H′	SR	SE	Den (ha^{-1})	Dbh (cm)	BA (m^2ha^{-1})	
Sacred groves	Osun-Osogbo	Nigeria	Intact	22	65	2.98	8.09	0.71	3078	24.4	32.69	
	Idanre Hills	Nigeria	Intact	26	60	3.17	7.11	0.77	4633	31.9	44.85	Onyekwelu et al. (2021a)
	Ogun-Onire	Nigeria	Intact	32	85	3.55	10.02	0.80	4825	28.0	40.70	
	Mawlong sacred	India	NA	22	41	3.22	NA	0.87	847	NA	32.3	Bdoor (2016)
	Konthoujam Lairembi	India	NA	NA	45	3.17	2.37*	0.83	NA	NA	NA	Khumbongmayum et al. (2005)
	Mahabali	India	NA	NA	20	2.48	1.21*	0.55	NA	NA	NA	
Natural Forests	Eda Forest Reserve	Nigeria	Degraded	17	28	2.58	5.71	0.45	301	40.7	40.68	Agbelade and Ojo (2020)
	Murtiya Forest	India	NA	20	23	2.37	4.27	0.76	172	34.4	71.73	

Balcha Forest	India	NA	10	11	1.35	2.19	0.56	96	32.2	58.46	Adekunle et al. (2014)
Nishangara Forest	India	NA	12	13	1.73	2.44	0.67	136	31.5	56.33	
Akure Enriched Forest	Nigeria	Enriched	21	42	3.25	7.58	0.87	446	23.3	35.20	Lawal and Adekunle (2013)
Akure Forest	Nigeria	Degraded	11	18	2.74	5.00	0.95	60	13.8	-	
Queen Forest	Nigeria	Primary	27	51	3.31	5.16	0.66	671	27.3	85.40	Onyekwelu et al. (2008)
Oluwa Forest	Nigeria	Degraded	24	45	3.12	5.04	0.60	513	22.5	35.90	
Elephant Forest	Nigeria	Degraded	22	31	2.82	4.98	0.57	508	21.6	29.40	

Where SR = Species Richness; SE = Species Evenness; NoF = Number of Families; NoS = Number of Species; BA - Basal Area;
* Species Richness index was used

Table 2: Comparison of Biodiversity Indices of Sacred Forests and Other Tropical Forest Types

Source: Onyekwelu, J.C. 2021b. "Can the Fear of the Gods Sustain Biodiversity Conservation in Sacred Groves?" *Academia Letters*, Article 635. Accessed July 6, 2022. https://doi.org/10.20935/AL635.

The report of high number of tree species in sacred groves compared to conventional tropical forests, further strengthens the general belief that sacred groves are important for *in-situ* biodiversity conservation. Generally, sacred groves harbour similar and/or higher species richness and diversity compared to natural (primary and degraded) and enrichment forests (Table 2). All sacred groves referenced on Table 2 have superior biodiversity indices compared to degraded forests, with the exception of Mahabali sacred grove in India. Many studies in Africa reported higher species diversity in sacred groves than adjourning natural forests, in some cases more species were reported in sacred groves than government-protected areas in similar regions. For example, sacred groves in Tanzania had higher woody species richness than state managed forests (Mgumia and Oba 2003). Onyekwelu and Olusola (2014) and Onyekwelu (2021a) observed that sacred groves in south-western Nigeria had comparable and/or higher species diversity than primary and degraded forests, respectively. Using the study of small mammal communities on the Accra Plains of Ghana, Decher (1997) established evidence for the ecological value of sacred groves. In Togo, Kokou et al. (2005; 2008) found 15 species that are new to science from their studies in sacred forests. Other studies have reported high biodiversity in sacred natural sites in many countries in Africa, among which are: Cote d'Ivoire (Ibo 2005); Ghana (O'Neal-Campbell 2005); Mozambique (Virtanen 2002); Morocco (Deil et al. 2005); etc.

Evidence of sacred groves serving as reservoir and repositories of rare and endemic species abound (Jamir and Pandey 2003; Kokou et al. 2008; Onyekwelu 2021b). The number of regenerating rare/endemic plant species was higher in some of the sacred groves compared to reserved forests (Boraiah et al. 2001). According to Kokou et al. (2008), sacred forests in Togo protect rare forest types, contain high biodiversity, and harboured threatened species found nowhere else in the country. In Nigeria, Onyekwelu and Olusola (2014) observed that about 32% of tree species found in Osun-Osogbo and Igbo-Olodumare sacred groves are among the endangered tree species in Nigeria. *Hildegardia barteri* and *Lovoa trichilioides* were noted to be endemic and unique to Igbo-Olodumare and Ogun-Onire sacred groves, respectively in south-western Nigeria (Onyekwelu et al. 2021). The presence of these two species has not been reported in any forest reserve in Nigeria (Onyekwelu et al. 2021). Also, some tree species that are not found in protected forests in Nigeria were encountered in some of the sacred groves. The prevalence of the unique species clearly indicates the potential of sacred groves for conservation of important tree species.

Going by the higher number of seedlings than saplings and overstory species in four sacred groves in Nigeria, Onyekwelu et al. (2021) concluded that sacred groves have potentials for conservation of regenerants, which is similar to the opinion of Ballabha et al. (2013), who noted that regeneration is good if the number of seedlings > number saplings > number of overstory trees. For example, Onyekwelu et al. (2021) reported 39–78 tree species at the understory layers of four sacred groves compared to the 32–62 tree species in the overstory layer, which is a confirmation that the understory of sacred groves could be more species rich than the overstory layer. The good regeneration status was further confirmed by the good species recruitment in the understory. Another indication of good species' regeneration in sacred groves is the higher tree density in the understory layer than the overstory layer, which is in consonance with some published results (Onyekwelu and Olusola 2014; Johnston 2019; Onyekwelu et al. 2021). Apart from species level biodiversity conservation, sacred groves are known to conserve ecosystems. For example, it has been observed that in some instances, sacred groves are the only remaining natural forests outside protected areas and that they are known to conserve habitats that are not represented in protected area systems (Bhagwat and Rutte 2006). Sacred forests have been reported to protect rare forest types (Kokou et al. 2008). They were the only remnants of dry evergreen forests and tropical afromontane forests in India and Ethiopia, respectively (Ramanujam and Cyril 2003; Aerts et al. 2006).

Ecosystem Services Provision

Sacred groves can provide important ecosystem services for local communities such as provisioning, regulating, supporting and cultural amenities (Ormsby 2012; Blicharska et al. 2013; Onyekwelu and Olusola 2014; Adeyanju 2020). Rich biodiversity and a wide array of cultural practices that take place in sacred groves suggest that they have potentials to provide concrete ecosystem services and less tangible cultural services to indigenous people. Ecosystem services are defined to include, provisioning services such asfood andwater; regulating services such as flood and disease control; cultural services such as spiritual, recreational,andculturalbenefits; and supporting services, such as nutrient cycling; that maintain the conditions forlifeonearth. However, ecosystem services are often considered intangible, subjective and difficult to measure and are therefore often undervalued or overlooked (Reuben and Kquofi 2015). The various ecosystem services provided by sacred groves are potentials for driving their protection (Onyekwelu and Olusola 2014; Adeyanju 2020; Onyekwelu

2021a; Sharma and Kumar 2021). Adeyanju (2020) observed that various religio-cultural benefits are the primary motivations to preserve sacred groves in south-western Nigeria. There is therefore no doubt that ensuring the sustainability of the numerous ecosystem services provided by sacred grove will contribute to its conservation and future relevance.

A very important ecosystem service provided by most sacred groves is the religious and spiritual function, since this is the primary reason for their establishment. To this end, most sacred groves are dedicated to or associated with deity(ies) that are usually worshiped by the people. This is especially true in Africa and Asia, where traditional systems of worship is still being practiced. Currently, many sacred groves function as places of worship, homes of deities, and preservation of culture. According to Sen (2018), the propitiation of deities in sacred groves has economic pertinence in terms of good rainfall, prosperous agricultural production, welfare of domestic animals and source of medicine, timber, fruits etc. during scarcity. In Nigeria, the worshipers of the Osun goddess in Osun-Osogbo sacred grove believe that by bathing in Osun River and participating in the annual Osun festival, they will receive blessings from the goddess to conceive baby(ies) since the goddess is believed to be the goddess of fertility (Adeyanju, 2020). The importance of sacred groves in community life is usually felt through religious-cultural practices. According to Adeyanju (2020), most of the ecosystem services provided by Osun-Osogbo sacred grove in Nigeria are cultural, which is evident from the 30%, 15% and 15% of community inhabitants who indicated that healing, promotion of art and culture, protection and peace, are important ecosystem services derived from the grove.

In the highlands of Ethiopia, the local people rely on Church forests for provisioning of livestock feed, tree seedlings, fuelwood, honey, clean water and other essential ecosystem services such as shade, climate regulation, habitat for pollinators and spiritual values (Cardelús et al. 2012; Aerts et al. 2016). In addition to spiritual functions, Adeyanju (2020) listed erosion control, windbreak, soil conservation, maintenance of hydrological cycle, watershed protection, water purification, maintenance of water quality, tourism and provision of NTFPs as some of the ecosystem services provided by some sacred groves in Nigeria. Through tourism, Mona monkeys in sacred groves in the Volta region of Ghana are been preserved, jobs created for people who act as local guides (Ormsby 2012) and traders in souvenirs, cultural and religious items, thus increasing their commitment to protecting the sacred groves, which is similar to what obtains in Osun-Osogbo grove (Adeyanju 2020). The medicinal properties of plants are well known to the local inhabitants, which explains why the people have been using different medicinal plants from sacred groves to cure different diseases. The medical uses of species in sacred groves have been recorded in Sierra Leone (Lebbie and Gurries 1995) and Tanzania (Msuya and Kideghesho 2009).

The sacred grove has protective functions, it protects plants, animals, the ecosystem, the people and their cultures. Idanre Hills sacred grove is a habitat for bats that probably play an important role in maintaining regeneration through seed dispersal. In Ethiopia, the last remaining fragments of tropical Afromontane forests are found in Church forests. Sacred forests in Togo protect rare forest types and harbour threatened species that are not found anywhere in the country (Kokou et al. 2008). In Nigeria, sacred groves harboured high percentage endangered trees species (Onyekwelu and Olusola 2014). Sacred groves contain certain unique species as well as species that perform key functions in nutrient conservation and could play important role in the rehabilitation of the surrounding degraded landscapes (Ramakrishnan 1996; Onyekwelu et al. 2021). The belief by local inhabitants that the god(s) protect them from diseases and natural calamities and provide healing and miracles, serve as motivations for protecting sacred groves. About 20% of the inhabitants of communities neighbouring Osun-Osogbo sacred grove in Nigeria indicated that protection, healing and miracles were among the benefits they derive from the sacred grove. Sacred groves have been reported to protect several crop wild relatives in India, including varieties of mango, jamun, fig and rare rattan species (Swamy et al. 2003). Biodiversity is a major ecosystem service provided by sacred grove, which was discussed in details above.

Sacred groves contribute significantly to carbon sequestration and climate change mitigation. Thus, sacred grove serves as sink of atmospheric CO_2. There is evidence in published literature that sacred groves accumulate high biomass and carbon stock, which could be higher or comparable to those of protected areas or primary forests. Sacred groves can accumulate mean biomass of 194.01 Mg ha^{-1} to 335.78 Mg ha^{-1} and carbon stock of 97.0 Mg C ha^{-1} to 156.04 Mg C ha^{-1} (Dar, 2019; Devi et al. 2021). Onyekwelu et.al. (2022) reported a mean total biomass range of 87.8 t ha^{-1} to 231.86 t ha^{-1}, and carbon stock of 43.9 t ha^{-1} to 115.9 t ha^{-1} for four sacred groves in south-western Nigeria, which compared favourably and higher than those of a tropical rainforest and urban forests in Nigeria, respectively (Agbelade and Onyekwelu 2020; Oke et al. 2020).

Socio-Economic Importance of Sacred Groves

Socio-economic function is becoming an important factor for preservation of sacred groves by indigenous people. Sacred forests provide community inhabitants with their basic needs and therefore increases their food security, livelihood at subsistence and cash level, implying that sacred forests contribute to improving the livelihoods of the people. Employment generation, tourism, income generation, and collection of NTFPs are some of the emerging factors that will be important in future preservation of sacred groves and their biodiversity conservation potentials (Figure 5.3; Onyekwelu and Olusola 2014).

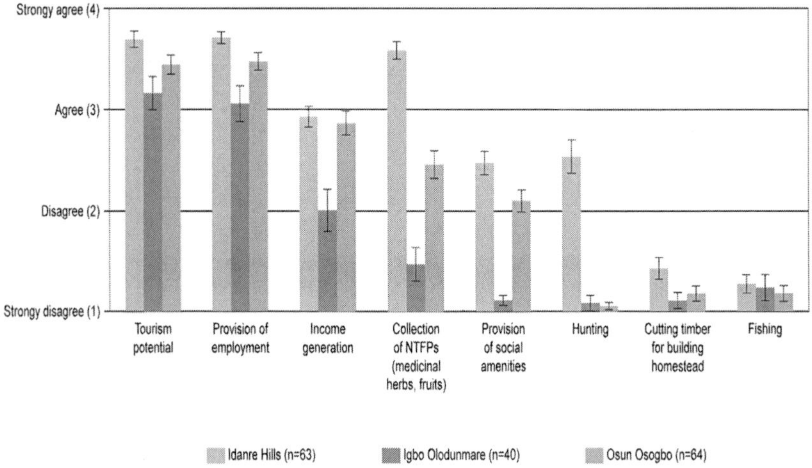

Figure 5.3: Socio-Economic Benefits Driving Biodiversity Conservation in Three Sacred Groves in South-Western Nigeria

Source: Adeyanju, S. O. 2020. "Drivers of Biodiversity Conservation in Sacred Groves: A Comparative Study of Three Sacred Groves in South-west Nigeria." M.Sc. Thesis, Canada: University of British Columbia. 98 pp. Accessed July 6, 2022. http://dx.doi.org/10.14288/1.0389686.

In many parts of the world, sacred groves are sites for practices such as rituals, initiations, festivals, and ceremonies, including the selection and installation of traditional rulers (Daniel et al. 2016), which has led to an alarming high influx of people into sacred groves. For example, it is estimated that between 130,000 and 150,000 people visit Osun-Osogbo sacred grove in Nigeria for annual Osun-Osogbo festival celebration (Aleshinloye and Maruyama 2015; Adeyanju 2020). In Tafi Atome Monkey

Sanctuary in Ghana, a festival to celebrate the monkeys take place every February, managed by the fetish priest and attended by villagers while at Boabeng-Fiema sanctuary, annual yam festival is held to celebrate the monkeys (Ormsby 2012). Rituals and the famous Orosun, Ogun and Ije festivals are some of the social activities that are frequently undertaken at Idanre Hills sacred grove in Nigeria. Upon coronation, the king of Idanre goes and stays on Idanre Hills sacred forests for some days. The sacred grove is also the burial site for the kings and some prominent Idanre Chiefs.

Employment is a key economic service provided by sacred groves. People are employed directly or indirectly in sacred groves. Currently, Osun-Osogbo sacred grove directly employs 73 people (most of whom are indigenous people), consisting of a curator, 27 security guards, 7 guides, 7 education officers, 3 masons, 3 traditional artists, 3 carpenters and 22 administrators (IUCN 2005). The Lopé-Okanda in Gabon is reported to employ 50 people (IUCN 2005). In addition to direct employment, sacred groves employ thousands of people indirectly, who are mostly undocumented. For example, people living in communities neighbouring sacred groves engage in the collection and sale of NTFPs, production and marketing of handicrafts, provision of hospitality and lodging for worshipers, visitors and tourists, provision of entertainment, sale of other goods and services during festivals and ceremonies. Udeagha et al. (2013), reported that a large proportion of the respondents (71%) generated income from various activities carried out within Asanting Ibiono sacred Forests in Nigeria, thus, contributing to their livelihood's economy. The communities neighbouring Igbo-Olodumare and Osun-Osogbo sacred groves benefited in several ways from the groves, with promotion of art and culture being the most prominent (50 and 40% of respondents in Igbo-Olodumare and Osun-Osogbo, respectively) (Onyekwelu and Olusola 2014). The hunting of bats during Ise festival in Idanre Hills sacred grove, Nigeria provides income for young men who take part in the hunting and selling of bats as well as helps to save money that would have been spent on meat (Adeyanju 2020).

Tourism and its accompanying benefits are other important socio-economic importance of sacred groves. While some sacred sites are not tourist destinations, tourism may represent a method to provide additional protection for sacred sites, including revenue to help with management and conservation (Ormsby 2012). Opening sacred groves to tourism can cause a shift in the cultural significance of the site, generate income, create employment and increase environmental impacts. Due probably to the awareness being created, increasingly more tourists are visiting sacred groves. Every year, up to 150,000 tourists visit Osun-Osogbo grove for the annual Osun festival celebration (Adeyanju 2020). In addition are thousands of other tourists who visit the sacred groves throughout the year. Onyekwelu and Olusola (2014) opined that the economic gains from tourism have increased the commitments of community dwellers to protect and preserve sacred groves. Tafifi Atome and Boabeng–Fiema Monkey Sanctuaries in Ghana received about 2,850 and 10,000 tourists in 2005, respectively (Ormsby 2012). Annual patronage of tourists at In Idanre Hills sacred grove was up to 8,700 in 2013 (Adigun et al. 2016). The development of ecotourism was suggested as an alternative means of motivating local communities to save the remaining patches of sacred forests in Zanzibar, Tanzania (Madeweya et al. 2004). About 48.1% of the residents of Boabeng and Fiema opined that they personally benefit from the tourism potentials of the sacred groves, some revenues from tourism at Tafifi Atome and Boabeng–Fiema Monkey Sanctuaries are shared with the communities (Ormsby 2012).

Potentials of Sacred Groves in Sustaining Provision of Important Values

Despite their numerous religious, cultural, social, economic and livelihood importance to indigenous people and the environment, sacred groves have, in recent times, come under strong anthropogenic pressures and threats through cultural changes, neglect, collapse of villages, modernization, decreasing resources, violation of rules/taboos, loss of sacredness leading to lack of fear of deity(ies), weakening or loss of traditional values/beliefs, encroachment, population growth, excessive visitors' population, changing socio-economic conditions, rural-urban migration, etc. (Dudley et al. 2010; Ormsby and Bhagwat 2010; Blicharska et al. 2013; Onyekwelu and Olusola 2014; Mouroufie et al. 2019; Adeyanju 2020; Onyekwelu 2021a&b; Sharma and Kumar 2021). Other factors include: decrease in the number of adherents and practitioners of ethnomedicine as well as fragmentation of families taking care of sacred groves, which may adversely affect their protection. Today, many sacred groves have shrunk to small areas while some have been abandoned. These threats have led to the alteration in size, vegetation structure, reduction in species composition, ineffectiveness of taboos, collapse of cultural and social systems and lack of protection of sacred groves. In some cases, the threats exceed the capacity of the sacred groves to maintain their integrity, which may lead to their outright destruction. There are indications that more sacred groves may lose their sacredness or be lost as the threats escalate; as the interest of young people in cultural traditions reduces making it difficult to replace ageing priest/

adherents; as belief systems changes thus reducing the fear/respect for the gods (Onyekwelu 2021b). The degradation or destruction of sacred groves has adverse effects on the indigenous people that depend on them and their environmental and religious systems. It threatens the long-term ability of sacred groves to conserve biodiversity and continue the provision of important ecosystem goods and services.

Thus, whether sacred groves can continue to provide their important ecosystem goods and services now and, in the future, depends largely on their status and management strategies. Whatever strategies are proposed or used must take into consideration the social, cultural, economic and spiritual needs of the people. According to Jhansi et al. (2018), sacred groves must be protected so that they can continue maintaining ecological and environmental balances and servicing local communities while Onyekwelu (2021b) canvased for government and outside support to indigenous people to protect sacred groves. Zannini et al. (2021) advocated for efforts towards filling knowledge gaps and improving broader knowledge of sacred natural sites to maximize their synergy with larger conservation schemes. Dudley et al. (2010) suggested that sacred grove system should be incorporated into broad conservation policies by governments and conservation organizations. The integration of sacred groves into conservation strategies may be an important way of preserving them as well as the biological and cultural values they represent. Onyekwelu (2021b) concluded that making Osun-Osogbo sacred grove a Nigerian National Monument and UNESCO world heritage site gave it legal backing, increased its protection, prevented degradation and improved its values, which has contributed in making Osun-Osogbo one of the best conserved and most popular sacred groves in Nigeria today.

Continued legal protection of sacred groves by government agencies in order to compensate for the weakening of traditional protection and degradation of the groves has been advocated (Gadgil and Vartak 1976; Dacher,1997). Communities seeking to protect their sacred groves should be recognized and empowered by the government to restrict activities that threaten their groves and to take legal action against those who encroach on the sites. Legislation and implementation of new laws and policies by government for management of sacred groves as well as creating awareness among people about the values of sacred groves are imperative (Soury 2007; Sharma and Kumar 2021). Government's involvement in promoting ecotourism in Idanre Hills sacred grove in Nigeria, including organizing mountain climbing, the annual musical and cultural Mare festival, were pivotal to the increased awareness and patronage of the grove while promoting the culture and traditions of Idanre people (Adigun et al. 2016). Adeyanju (2020) recommended institutionalization and collaborative management of sacred groves between government and traditional institutions.

As measures to harness the important roles of sacred groves, Onyekwelu (2021a) suggested three strategies: (i) identifying the drivers of conservation in sacred groves and strengthening them (ii) increasing protection of sacred groves and (iii) identifying the threats to sacred groves and addressing them. Vipat and Bharucha (2014) advocated for alternate conservation strategies, including: ecodevelopment, eco-restoration and sustainable tourism, through registering local Biodiversity Management Committees to prevent the destruction of the fragile and biologically rich ecosystems of sacred groves. Since it is obvious that the traditional role of sacred groves may inevitably fade away due to modernization and the younger generation may become more interested in modern cultural values and habits than in traditional customs and taboos (Decher 1997; Adeyanju 2020; Onyekwelu 2021a), Decher (1997) advocated for increased roles of NGOs in sacred grove management. Knowledge of the locations and boundaries of sacred forests and their detailed map are necessary for their protection and to increase their knowledge and awareness.

References

Adekunle, V.A.J, N.K. Nair, A.K. Srivastava, and N.K. Singh. 2014. "Volume Yield, Tree Species Diversity and Carbon Hoard in Protected Areas of Two Developing Countries." *Forest Science and Technology* 10(2): 89-103.

Adeyanju, S.O. 2020. "Drivers of Biodiversity Conservation in Sacred Groves: A Comparative Study of Three Sacred Groves in South-West Nigeria." M.Sc. Thesis, Canada: University of British Columbia. 98 pp. Accessed July 6, 2022. http://dx.doi.org/10.14288/1.0389686.

Adigun, F.O., F. Abolade, and D.I. Adegboye. 2016. "Patronage Pattern of Idanre Hills as Ecotourism Centre." *Research on Humanities and Social Sciences* 6(12): 35–42.

Aerts, R., K.Van Overtveld, M. Haile, M. Hermy, J. Deckers, and B. Muys. 2006. "Species Composition and Diversity of Small Afromontane Forest Fragments in Northern Ethiopia." *Plant Ecology* 187: 127–142.

Aerts, R., K. Van Overtveld, E. November, A. Wassie, A. Abiyu, S. Demissew, D.D. Daye, K. Giday, M. Haile, S. TewoldeBerhan, D. Teketay, Z. Teklehaimanot, P. Binggeli, J. Deckers, I. Friis, G. Gratzer,

M. Hermy, M. Heyn, O. Honnay, M. Paris, F. J. Sterck, B. Muys, F. Bongers, and J.R. Healey. 2016. "Conservation of the Ethiopian Church Forests: Threats, Opportunities and Implications for their Management." *Science of the Total Environment.* Accessed June 20, 2022. DOI: 10.1016/j.scitotenv.2016.02.034.

Agbelade, A.D. and B.H. Ojo. 2020. "Species Diversity, Volume Determination and Structure of Protected Forests for In-Situ Biodiversity Conservation." *International Journal of Conservation Science* 11(1): 133-144.

Agbelade, A.D. and J.C. Onyekwelu. 2020. "Tree Species Diversity, Volume Yield, Biomass and Carbon Sequestration in Urban Forests in Two Nigerian Cities." *Urban Ecosystems.* Accessed June 20, 2022. https://doi.org/10.1007/s11252-020-00994-4.

Ajbilou, R., T. Marañón, and J. Arroyo. 2006. "Ecological and Biogeographical Analyses of Mediterranean Forests of Northern Morocco." *Acta Oecologica* 29: 104–113.

Aleshinloye, D. and N. Maruyama. 2015. "Solidarity at the Osun Osogbo Sacred Grove, a UNESCO World Heritage Site." *J Tour Plan Dev* 13(3): 274–291.

Aniah, P. and A. Yelfaanibe. 2016. "Learning from the Past: The Role of Sacred Groves and Shrines in Environmental Management in the Bongo District of Ghana." *Environmental Earth Sciences* 75 (75): 916. Accessed May 15, 2022. https://doi.org/10.1007/s12665-016-5706-2.

Arora, V. 2006. "The Forest of Symbols Embodied in the Tholung Sacred Landscape of North Sikkim, India." *Conservation and Society* 4(1): 55–83.

Bdoor, Bshar Samir. 2016. "Tree Diversity and Population Structure in Sacred and Non-sacred Forests on Karst Landscape in Meghalaya, Northeast, India." *Int. J. Sci. Tech* 11: 17-32.

Ballabha, R., J.K. Tiwari, and P. Tiwari. 2013. "Regeneration of Tree Species in yhe Sub- Tropical Forest of Alaknanda Valley, Garhwal Himalaya, India." *For Sci Pract* 15: 89–97.

Ballullaya, U.P., K.S. Reshmi, T.P. Rajesh, K. Manoj, M. Rajesh, M. Lowman, and S.P. Allesh. 2019. "Stakeholder Motivation for The Conservation of Sacred Groves." In: South India: An Analysis of Environmental Perceptions of Rural and Urban Neighbourhood Communities." *Land Use Policy* 89: 104213.

Bhagwat, S.A. and C. Rutte. 2006. "Sacred Groves: Potential for Biodiversity Management." *Frontiers in Ecology and the Environment* 4(10): 519–524. Accessed July 6, 2022. http://doi.org/10.1890/1540-9295-(2006)4 [519: SGPFBM]2.0.co;2.

Blicharska, M., G. Mikusiński, A. Godbole, and J. Sarnaik. 2013. "Safeguarding Biodiversity and Ecosystem Services of Sacred Groves – Experiences from Northern Western Ghats." *International Journal of Biodiversity Science, Ecosystem Services and Management.* 9: 4, 339-346. Accessed July 6, 2022. DOI: 10.1080/21513732.2013.835350.

Boadi, S., C.A. Nsor, D.H. Yakubu, E. Acquah, and O.O. Antobre. 2017. "Conventional and Indigenous Biodiversity Conservation Approach: A Comparative Study of Jachie Sacred Grove and Nkrabea Forest Reserve." *International Journal of Forestry Research*, Vol. 2017, 8 pp. Accessed November 9, 2021. https://doi.org/10.1155/2017/1721024.

Boraiah, K.T., R. Vasudeva, S. A. Bhagwat and C. G. Kushalappa. 2003. "Do informally Managed Sacred Groves Have Higher Richness and Regeneration of Medicinal Plants than State-Managed Reserve Forests?" *Current Science* 84(6): 804–808.

Borokini T.I. 2016. "Sanctuary of the Spirits: Okwu-muo, Ori Oke and 'Mammy Water' in the Veneration of Sacred Natural Sites in Southern Nigeria." *International Journal of Intangible Heritage* 11: 56 – 70. Accessed March 11, 2022. http://doi.org/10.35638/ijih.2016.11.006.

Chukwu, O., C.I. Ezeano, J.U. Ezenwenyi, and M.A. Adeyemi. 2019. "Impact of Cultural Belief on the Preservation of Agunabani Sacred Forest in Okposi, Nigeria." *Research Journal of Forestry* 13: 9-13.

Cardelús, C.L., M.D. Lowman, and A.W. Eshete. 2012. "Uniting Church and Science for Conservation." *Science* 335: 915-916.

Daniel, K.S., A.U. Udeagha, and D.E Jacob. 2016. "Horticulturalist Importance of Sacred Forests Conservation in South Southern Nigeria." *African Journal of Sustainable Development* 6(2): 251–268.

Dar, J.A., K. Subashree, D. Raha, A. Kumar, P.K. Khare, and M.L. Khan. 2019. "Tree Diversity, Biomass and Carbon Storage in Sacred Groves of Central India." *Environmental Science and Pollution Research.* Accessed June 10, 2022. https://doi.org/10.1007/s11356-019-06854-9.

Davi, N.B., N.T. Lepcha, S.S. Mahalik, D. Dutta, and B.L. Tsanglao. 2021. "Urban Sacred Grove Forests are Potential Carbon Stores: A Case Study from Sikkim Himalaya." *Environmental Change* 4 (100072). Accessed May 15, 2022.
https://doi.org/10.1016/j.envc.2021.100072.

Decher, J. 1997. "Conservation, Small Mammals and the Future of Sacred Groves in West Africa." *Biodiversity and Conservation* 6: 1007–1026.

Deil, U., H. Culmsee, and M. Berriane. 2005. "Sacred Groves in Morocco: A Society's Conservation of Nature for Spiritual Reasons." *Silva Carelica* 49: 185–201.

Dudley, N., S. Bhagwat, L. Higgins-Zogib, B. Lassen, B. Verschuuren, and R. Wild. 2010. "Conservation of Biodiversity in Sacred Natural Sites in Asia and Africa: A Review of the Scientific Literature." In: B. Verschuuren, R. Wild, J. McNeely, and G. Oviedo, G. eds. *Sacred Natural Sites: Conserving Nature and Culture*. London EC1N 8XA, UK: Earthscan. Accessed July 6, 2022. DOI: https://doi.org/10.4324/9781849776639.

Emeasoba, U.R.B. and J.U. Ogbuefi. 2013. "The Dynamics of Land Ownership by Deities in Anambra State Nigeria." *Research on Humanities and Social Sciences* 3(7): 61-67.

Gadgil, M. and V.D. Vartak. 1975. "Sacred Groves of India: A Plea for Continued Conservation." *Bombay Nat Hist Soc.* 72: 313–20.

Gadgil, M. and V.D. Vartak. 1976. "The Sacred Groves of Western Ghats in India." *Econ. Bot.* 30: 152-60.

Ibo, J. 2005. "Contribution des organisaiton non gouvernmentales ecologistes a l'amenagement des forest sacreses en Cote d'Ivoire: L'experience de la Croix Verte, Vertigo. *La revue en sciences de l'environnement* 6(1): 1–13.

Irvine, K.N., D. Hoesly, R. Bell-Williams, and S.L. Warber. 2019. "Biodiversity and Spiritual Well-being." In: *Biodiversity and Health in the Face of Climate Change*. M. R. Marselle et al. Eds. Springer Nature Switzerland, 213 - 247. Accessed July 6, 2022. https://doi.org/10.1007/978-3-030-02318-8_10.

IUCN. (International Union for Conservation of Nature). 2005. Osun-Osogbo Sacred Grove (Nigeria), ID no. 1118. In: *Evaluations of Cultural Properties*. UNESCO World Heritage Convention, 31–38. Accessed August 11, 2022. http://unesdoc.unesco.org/images/0014/001488/148831e.pdf.

Jamir, S.A. and H.N. Pandey. 2003. "Vascular Plant Diversity in the Sacred Groves of Jaintia Hills in Northeast India." *Biodiversity and Conservation* 12: 1497–1512.

Jhansi, K., S.M. Khasim and R.M. Venkatesh. 2018. "Ethnobotanical Medicinal Plants in ParammaKonda, Konada, Nareduvalsa and Jayathi Sacred Groves, Vizianagaram, Andhra Pradesh." *India Ann Plant Sci* 7(3): 2127–2140.

Johnston, J. 2019. Overstory and Understory Species Composition and Diversity in UNDERC Closed-canopy Hardwood Forests BIOS 35502-01: practicum in environmental field biology. Accessed June 10, 2022. https://underc.nd.edu/assets/338767/fullsize/johnston_2019.pdf.

Khan, M.L., D.V. Khumbongmayum, and R.S. Tripathi. 2008. "The Sacred Groves and their Significance in Conserving Biodiversity—An

Overview." *International Journal of Ecology and Environmental Sciences* 34: 277–291.

Kidegheso, J. 2008. "Co-existence between the Traditional Societies and Wildlife in Western Serengeti, Tanzania: Its Relevancy in Contemporary Wildlife Conservation Efforts." *Biodiversity Conservation* 17: 1861–1881.

Kokou, K., K. Adjossou, and K. Hamberger. 2005. *Les forêts sacrées de l'aire Ouatchi au sud-est du Togo et les contraintes actuelles des modes de gestion locale des ressources forestieres. Revue electronique Vertigo* 6(3), 9 pp.

Kokou, K., K. Adjossou, and A.D. Kokutse. 2008. "Considering Sacred and Riverside Forests in Criteria and Indicators of Forest Management in Low Wood Producing Countries: The Case of Togo." *Ecological Indicators* 8: 158–169.

Khumbongmayum, A.D., M.L. Khan and R.S. Tripathi. 2005. "Sacred Groves of Manipur, Northeast India: Biodiversity Value, Status and Strategies for their Conservation." *Biodiversity and Conservation* 14: 1541–1582. Accessed March 5, 2022. DOI 10.1007/s10531-004-0530-5.

Laird, S.A. 1999. Forests, Culture and Conservation. In: D.A. Posey. Ed. *Cultural and Spiritual Values of Biodiversity. A Complementary Contribution to the Global Biodiversity Assessment,* 347– 396. Nairobi: UN Environment Programme.

Lawal, A. and V.A.J. Adekunle. 2013. "A Silvicultural Approach to Volume Yield, Biodiversity and Soil Fertility Restoration of Degraded Natural Forest in South-West Nigeria." *Inter. Journal of Biodiversity Science Ecosystem Services and Management* 9(3): 201-214.

Lebbie, A.R. and R.P. Guries. 1995. "Ethnobotanical Value and Conservation of Sacred Groves of the Kpaa-Mende in Sierra-Leone." *Economic Botany* 49: 297–308.

Madeweya, K.H., H. Oka, and M. Mat. 2004. "Sustainable Management of Sacred Forests and their "Potential for Eco-tourism in Zanzibar." *Bulletin of FFPRI* 1(390): 33–48.

Mgumia, F.H. and G. Oba. 2003. "Potential Role of Sacred Groves in Biodiversity Conservation in Tanzania." *Environ Conserv* 30: 259–265.

Mouroufie, K.K.V., M.V. Koffi, B. OgniKanga, A. Bitty, and B.S. Gonedele. 2019. "Socio-Anthropological Factors as Sources of Desecration of Grébouo 1 Sacred Grove, Southwestern Cote d'Ivoire." *European Scientific Journal* 15(5): 1857 – 7881.

Mukaria, A. 2021. "Healing through Nature. A Lesson from the Traditional Ameru Culture in Kenya." *Academia Letters,* Article 1350. Accessed July 6, 2022. https://doi.org/10.20935/AL1350.

Myers, N., R.A. Mittermeier, C.G. Mittermeier, G.A. Da Fonseca, and J. Kent. 2000. "Biodiversity Hotspots for Conservation Priorities." *Nature* 403: 853–858.

Msuya, T.S. and J.R. Kideghesho. 2009. "The Role of Traditional Management Practices in Enhancing Sustainable Use and Conservation of Medicinal Plants in West Usambara Mountains, Tanzania." *Tropical Conservation Science* 2: 88–105.

National Commission for Museums and Monuments (NCMM). 2007. Oke Idanre (Idanre Hills). Nomination to the World Heritage List of Oke Idanre (Idanre Hills), Ondo State, Nigeria. Accessed May 2, 2022. https://whc.unesco.org/en/tentativelists/5169/.

Oke, O.S., S.O. Akindele, and J.C. Onyekwelu. 2020. "Biomass and Carbon Stock Assessment of a Tropical Rain Forest Ecosystem in Nigeria." *Journal of Forestry Research and Management* 17(2): 83-92.

Orlowska, I. and P. Klepeis. 2018. "Ethiopian Church Forests: A Socio-religious Conservation Model under Change." *Journal of Eastern African Studies* 12(4): 674-695.

O'Neal-Campbell, M. 2005. "Sacred Groves for Forest Conservation in Ghana's Coastal Savannas: Assessing Ecological and Social Dimensions." *Singapore Journal of Tropical Geography* 26: 151–169.

Onyekwelu, J.C., R. Mosandi, and B. Stimm. 2008. "Tree species diversity and soil status of Primary and Degraded Tropical Rainforest Ecosystems in South-Western Nigeria." *Journal of Tropical Forest Science* 20(3): 193–204.

Onyekwelu, J.C. and A. Olusula. 2014. "Role of Sacred Grove in In-Situ Biodiversity Conservation in Rainforest Zone of South Western Nigeria." *Journal of Tropical Forest Science* 26(1): 5–15.

Onyekwelu, J.C. 2021a. "Significance of Traditional Forest Management System to Biodiversity Conservation and Ecosystem Services Provision: Case study of Osun-Osogbo Sacred Grove, Nigeria." In: Felbermeier, Stimm and Seidl. eds. *Waldbau weltweit 2.0. Festschrift anlässlich der Verabschiedung von Prof. Dr. Dr. Reinhard Mosandl.* Munich, Germany: Forstliche Forschungsberichte 219:132 – 148.

Onyekwelu, J.C. 2021b. "Can the Fear of the Gods Sustain Biodiversity Conservation in Sacred Groves?" *Academia Letters*, Article 635. Accessed December 12, 2021. https://doi.org/10.20935/AL635.

Onyekwelu, J.C., A. Lawal, R. Mosandl, B. Stimm, and A. D. Agbelade. 2021. "Understory Species Diversity, Regeneration and Recruitment

Potential of Sacred Groves in South West Nigeria." *Tropical Ecology* 62: 427–442. Accessed December 12, 2021. https://doi.org/10.1007/s42965-021-00157-2.

Onyekwelu, J.C., A.D. Agbelade, B. Stimm, and R. Mosandl. 2022 (in press). "Role of Sacred Groves in Southwestern Nigeria in Biodiversity Conservation, Biomass and Carbon Stock Accumulation." (Manuscript under Consideration for Publication by Southern Forests: A Journal of Forest Science).

Ormsby, A. 2012. "Cultural and Conservation Values of Sacred Forests in Ghana." In: *Sacred Species and Sites: Advances in Biocultural Conservation*. Edited by G. Pungetti, G. Oviedo and D. Hooke. Cambridge: Cambridge University Press, pp. 335 - 350.

Ormsby, A.A. and S.A. Bhagwat. 2010. "Sacred Forests of India: A Strong Tradition of Community-Based Natural Resource Management." *Environmental Conservation* 37 (3): 320–326.

Parthasarathy, N. and K.N. Babu. 2019. "Sacred Groves: Potential for Biodiversity and Bioresource Management." In: W.L. Filho et.al. eds., *Life on Land, Encyclopedia of the UN Sustainable Development Goals,* Springer Nature. Switzerland AG. 15 pp. Accessed July 6, 2022. doi.org/10.1007/978-3-319-71065-5_10-1.

Probst, P. 2013. "Preserving Heritage and the Values of Exchange: Lessons from Nigeria." *History Compass.* 11(12): 1035–1046. Accessed May 15, 2022. https://doi.org/10.1111/hic3.12102.

Ramakrishnan, P.S. 1996. "Conserving the Sacred: From Species to Landscapes." *Nature and Resources* 32(1): 11–19.

Ramanujam, M.P. and K.P.K. Cyril. 2003. "Woody Species Diversity of Four Sacred Groves in the Pondicherry Region of South India." *Biodiversity and Conservation* 12: 289–299.

Reuben, G. and S. Kquofi. 2015. "Direct and Indirect Conservation Benefits of Religious and Spiritual Beliefs in Traditional Ghanaian Communities." *International Journal of Development and Sustainability* 4(3): 322-336.

Rutte, C. 2011. "The Sacred Commons: Conflicts and Solutions of Resource Management in Sacred Natural Sites." *Biological Conservation* 144: 2387–2394. Accessed March 5, 2022. https://doi.org/10.1016/j.biocon.2011.06.017.

Saj, T.L., C. Mather, and P. Sicotte. 2006. "Traditional Taboos in Biological Conservation: the Case of Colobus vellerosus at the Boabeng- Fiema Monkey Sanctuary, Central Ghana." *Social Science Information* 45(2): 285-310.

Salick, J., A. Amend, D. Anderson, K. Hoffmeister, B. Gunn, and F. Zhendong. 2007. "Tibetan Sacred Sites Conserve Old Growth Trees and Cover in the Eastern Himalayas." *Biodiversity and Conservation* 16: 693. Accessed May 15, 2022. https://doi.org/10.1007/s10531-005-4381-5.

Sen, U.K. 2018. "Assessing the Social, Ecological and Economic Impact on Conservation Activities Within Human-Modified Landscapes: A Case Study tn Jhargram District of West Bengal, India." *International Journal of Conservation Science* 9(2): 319-336.

Schlee, G. 1992. "Ritual Topography and Ecological Use: The Gabbra of the Kenyan/Ethiopian Borderlands." In: *Bush Base: Forest Farm, Culture, Environment Development.* Edited by E. Croll and D. Parkin. New York, USA, pp. 110–130.

Sharma, S, and Kumar, R. 2021. Sacred Groves of India: Repositories of a Rich Heritage and Tools for Biodiversity Conservation. *J. For. Res*. 32: 899–916. Accessed December 12, 2021. https://doi.org/10.1007/s11676-020-01183-x.

Shepherd, Gill. 1992. Managing Africa's Tropical Dry Forests: A Review of Indigenous Methods. *ODI Agricultural Occasional Paper* 14, London, UK: Overseas Development Institute. 117 pp.

Swamy, P.S., M. Kumar, and S.M. Sundarapandian. 2003. "Spirituality and Ecology of Sacred Groves in Tamil Nadu." *Unasylva* 54: 53–58.

Soury, A. 2007. *Sacred Forests: A Sustainable Conservation Strategy? The Case of Sacred Forests in the Ouémé Valley*, Benin. M.Sc. Thesis, Wageningen University, the Netherlands: Department of Social Sciences, 142 pp. Accessed May 4, 2022. https://edepot.wur.nl/199200.

Udeagha, A.U., S.I. Udofia, and D.E. Jacob. 2013. "Cultural and Socio-economic Perspectives of the Conservation of Asanting Ibiono Sacred Forests in Akwa Ibom State, Nigeria." *International Journal of Biodiversity and Conservation* 5(11): 696-703. Accessed March 5, 2022. DOI: 10.5897/IJBC2013.0580.

Vipat, A. and E. Bharucha, 2014. "Sacred Groves: The Consequence of Traditional Management." *Journal of Anthropology*. Vol. 2014. 8pp. Accessed May 15, 2022. http://dx.doi.org/10.1155/2014/595314.

Virtanen, P. 2002. "The Role of Customary Institutions in the Conservation of Biodiversity: Sacred Forests in Mozambique." *Environmental Values* 11(2): 227–241.

Yusuf, T.G. 2016. "A Micro Analysis of Tourists, Other Participants and Tourism Activities at Osun-Osogbo Sacred Grove, Nigeria." *Journal of Economics and Sustainable Development* 7(7): 96–104.

Zannini, P., P. Frascaroli, J. Nascimbene1, A. Persico, J.M. Halley, K. Stara, G. Midolo, and A. Chiarucci. 2021. "Sacred Natural Sites and Biodiversity Conservation: A Systematic Review." *Biodiversity and Conservation* 30: 3747–3762. Accessed June 10, 2022. https://doi.org/10.1007/s10531-021-02296-3.

CHAPTER VI

CAPITAL CONVERSION INSTITUTIONS: SACRED GROVES IN NORTHERN TANZANIA

MICHAEL J. SHERIDAN

Ever since James Frazer based *The Golden Bough* (1890), his monumental survey of religion and ritual, on the theme of holy trees, Westerners have glossed these areas as "sacred groves" and perceived them as icons of tradition and order in small-scale societies. Although Africanist scholars had long depicted sacred trees, groves, and forests as "ethnographic curiosities" (Castro 1990, 277), more recent work has recast them as resource management systems with the potential to conserve biodiversity and mitigate deforestation in Africa (Byers et al. 2001; Dudley et al. 2005). A 2005 symposium on the management of sacred sites called for administrators, scholars, and activists to work toward "safeguarding the cultural and biological diversity embodied in sacred natural sites and cultural landscapes" (UNESCO 2005), efforts that surely would contribute much to the diversification of conservation policy. Yet much of the conservation policy literature for Africa treats natural sacred sites as relics of ostensibly pristine ecosystems and unchanged traditional values (e.g., Decher 1997; Tankou et al. 2014). Such ecological and cultural stasis will surely be scarce in Africa under conditions of global capitalism and climate change, and the assumption that sacred groves themselves are static entities may lead to inappropriate conservation policies and unexpected outcomes. The ecological implications of ritual may have the potential to enhance African land use, but the functionalist assumption that sacred groves exist primarily as indigenous forms of conservation (e.g., Chidhakwa 2003) is mistaken. This case study from Tanzania rebuts this "relic theory" and asserts that sacred groves are sites where ecological, social, political, and cultural dynamics intersect and disconnect over time. The evidence from a farming community in the Tanzanian highlands suggests that sacredness is a contested process among unequal social actors, not a universally shared and consistently practiced specialness. The process-based approach described here suggests that conservation policy for sacred groves should be oriented not simply toward empowering locals to protect and maintain these sites, but rather toward adaptive co-management by various institutions at local, national, and global levels.

Sacred groves exist throughout tropical Africa and typically serve as places for rituals of initiation and sacrifice. They usually consist of dense patches of forest in agrarian landscapes, and are particularly common in the long arc of forest–savanna transition zone from West Africa to Southern Africa (Sheridan 2008; 2016). The term *sacred* does not imply, however, that these sites are purely religious institutions, separate from politics, social organization, and land tenure—and effective conservation policy should not make such an assumption. African sacred groves are "places of power" (Colson 1997), and as such they are critical sites in the ideological and material struggles that generate political legitimacy, ethnic and gender identities, and access to resources. They are institutional contexts for processes of landscape formation and the meaningful construction of material space as culturally significant place. Instead of being forested remnants of social and ecological continuity, African sacred groves make social institutions manifest, and this is why they are important places for social action and cultural change. This chapter assesses the ecological status of sacred groves in an East African mountain block, describes their social history, and suggests strategies for their conservation in Tanzania and, by extension, throughout Africa.

African Sacred Groves and Conservation

Postcolonial conservationists began to focus on sacred groves only recently. In the 1990s the failures of state-led conservation efforts in Africa led policy-makers to create less centralized, community-based institutions for natural resource management (Adams and Hulme 2001). Arguing that the taboos protecting sacred groves constitute indigenous conservation systems (Colding and Folke 2001), they searched for ways to incorporate these practices into formal conservation policy (e.g., Dorm-Adzobu et al. 1991; Gerdén and Mtallo 1990; Robertson and Luke 1993) (1). Various international NGOs sponsored community-based projects in African sacred groves (Anane 1997; Nyamweru 1996; Schaaf 1998; UNDP 2004), while scholars insisted that further attention to sacred groves was essential for building effective conservation and development (e.g., Awuah-Nyamekye 2009; Cardelús et al. 2019; Lebbie and Freudenberger 1996; Madeweya et al. 2004). The threatened status of African sacred groves has been recognized since the continent's colonial days, and recognition of this threat forms part of the standard argument in favor of their conservation (2). As Chevalier (1933, 37) phrased it, "since our contact the primitive has abandoned his creed, and the sacred groves are disappearing." Indeed, the trees in these groves—and the social institutions built upon them—face serious threats from local demand for fuelwood and global demand for forest products. In many areas, axes and chainsaws have replaced sacrifices and initiation ceremonies. The standard argument therefore asserts that because the groves are relics of primeval forests and the property of kin and/or ethnic groups with "traditional" spiritual values that conserved them, the appropriate policy is to empower and validate traditional institutions, practices, and beliefs.

Although persuasive, this argument is flawed, particularly in its assumption that African sacred groves are remnants of an ecological and cultural climax. Recent scholarship has questioned the "naturalness" of tropical forests around the world (Willis et al. 2004). Archaeologists have shown, for example, that the largest contiguous tract of "virgin" rainforest in the southern Amazon had been a thoroughly transformed "cultural parkland" before European contact (Heckenberger et al. 2003), and many of the "forest islands" in West Africa's savanna-forest transition zone are anthropogenic as well (Fairhead and Leach 1996). Indeed, this analytical shift from a focus on static entities to an emphasis on historical processes in tropical ecology has parallels in Africanist history, anthropology, and conservation policy. Current African religious practices, even those declared "traditional," are now understood not in terms of precolonial patterns but rather as recombinations, reinventions, and reconfigurations of belief systems (Mbiti 2015; Ranger 1988). The notion that "tribes" and ethnic groups are basic units of African societies has been replaced by attention to the construction of shifting identities in various historical contexts (e.g., Bravman 1998; Comaroff and Comaroff 2009). Finally, scholars have shown how community-based conservation often leads to institutional ambiguity and injustice rather than effective management (Larson and Ribot 2005; Martin et al. 2015). These intellectual currents suggest that the focus is shifting from fixed units and closed systems to dynamic and historically contingent complexities. Clearly, the assumption that sacred sites represent an unproblematic precolonial-era balance of nature and culture is inadequate for building effective conservation policy today.

Hybridized analytical approaches for documenting the dynamic nature of African sacred groves are just beginning (Courmier-Salem and Bassett

2007; Elliott and Campbell 2002; Juhé-Beaulaton 2010; Juhé-Beaulaton and Salpeteur 2018; Klepeis et al. 2016). Soil mechanics must meet institutional analysis and cultural meaning to account for these sites' complexities. Some groves may be relics of larger forests, but a more dynamic view of ecology and its linkages with social organization, politics, and symbolism shows that many are the historically contingent products of vegetation change and shifting social values and functions. Many, in fact, have been found atop old settlement sites, despite the current caretakers' insistence that the trees are relics of primeval forest (Chouin 2002; Fairhead and Leach 1997, 12; Guillot 1980; Lebbie and Guries 2008; Nyamweru et al. 2008), and many acquired spiritual value only within the context of—and possibly in reaction to—the transformations of colonial rule (Baum 1999; Campbell 2004, 227, Greene 2002). African sacred groves are community institutions in which power and authority are local in space and temporally tied to the life-cycles of kin groups and social networks. The people who care for them range from a few members of a patrilineage segment to kings and their coteries (Sheridan 2016). They often represent legitimate resource ownership and political influence, which accounts for much of their persistence compared to other forest areas throughout the postcolonial era in Africa. Yet it is often the groves' poor fit with the secular and religious institutions of the contemporary nation-state that threatens their destruction. The sacredness of African groves, then, is not embedded in their trees and plants, but rather in the diverse social institutions that sacred groves manifest in particular African landscapes. An analytical focus on intersecting ecological, social, and historical dynamics therefore offers a framework for understanding African sacred groves that avoids assumptions of steady states of both habitat and culture.

The Dynamics of Sacred Groves

Sacred groves are "natural cultural sites" (Nyamweru 2012) that typically serve as historical markers and sites for initiation, burial, and sacrifice in horticultural and agricultural societies (3). Igor Kopytoff's model of the "internal African frontier" (1988) can help us make sense of the remarkable similarities and continuities among sacred groves across tropical Africa. Kopytoff argues that although land was the scarce factor of production in Eurasian societies, in sparsely populated Africa labor was scarce. African societies therefore developed ways to organize rights in people, rather than in land, along moving frontiers of population growth. Kopytoff suggests the following growth cycle for precolonial African societies. First, some members of a mature "core" society moved to its periphery due to internal pressures or external opportunities. These people built a new society by recombining or innovating upon the institutions, practices, and symbols of the core society. The new society grew and devised ways to attract and incorporate new settlers into the community. As the population grew, elites emerged and shifted social organization from a focus on kin groups to the more organic solidarity of rulers and subjects. Some people left the new core society, and the process began again.

According to Kopytoff, this model explains the wide geographic spread of African cultural characteristics such as kinship, age, and gender hierarchies, sacrifice, initiation, and the political primacy of "firstcomer" groups over subsequent immigrants. Sacred groves were a key mechanism for reconstituting societies on these agrarian frontiers, and this may account for the relative lack of sacred groves among Africa's foraging and pastoralist societies. Sacred trees and groves marked new settlements, provided sites for the ritual practices that organized society, and served as icons of political legitimacy, prosperity, and health. If African societies were, as Kopytoff says, in "ceaseless flux" (1988, 7) before European colonization, the multiple significances of sacred grove institutions (in ecological, sociopolitical, and moral domains) provided much of that flexibility. This explains the presence of sacred groves across Africa in societies with quite different degrees of political centralization.

Some vocabulary from social theory can help to illuminate this process. Pierre Bourdieu argued that cultural meaning and social institutions are the results of people playing a grand game of interaction (1986). He refers to the counters for all of this strategizing as types of "capital" that actors seek to exchange and accumulate. Economic capital in the forms of land and money may indeed make the world go 'round, but people also tend to be occupied by gaining social and political capital (group memberships and patron-client networks), cultural capital (skills, status markers), and symbolic capital (honor, respect, significance). How, then, does a person with money convert that into social status and prestige? They could purchase and display something valuable but impractical, such as the gold wedding ring on my finger. Or they could sponsor a ritual in a sacred grove to communicate with their ancestors, which then gives them a network of co-participants and some recognition as a person of high moral character. Sacred groves are, in this sense, capital conversion machines in agrarian landscapes. This approach helps to avoid the unquestionability of sacredness and the all-too-easy assumption that the holy persists outside of history. It also allows an analyst to discuss African sacred groves in the sort of social science vocabulary that communicates across disciplines and topical specialization to capture the interest of policy makers steeped in institutional economics and cost-benefit analysis (Galiani and Sened 2014; North 1990).

The continuities of sacredness are based in the usefulness of sacred groves as homes for various institutions, not any essential traditionalism in African societies. The institutions embedded in the groves relate to critical factors of production such as access to land and control of labor, so the social context of sacred groves can have direct effects on their ecology. In many parts of Africa, sacred groves represent the moral authority of male ancestors, and are therefore material expressions of the social organization of society by gender and kinship (Fortes 1945, 101; Netting 1969, 1044). Sacred groves reflect idealized models of social order, and the ongoing struggle to create order makes them sites of cooperation and conflict instead of centers of static tradition. The political aspects of sacred groves are particularly cogent in frontier societies, especially those characterized by a division of rights and labor between autochthonous first settlers and later immigrants. The following case study shows how sacred groves assert the political primacy and authenticity of a group claiming to have been an area's first settlers. As markers of social differences that often relate to key institutions, sacred groves, as well as other sacred sites in Africa, function as political arenas in different historical contexts (Ellis and ter Haar 2004; Ranger 1999). These institutional and cultural linkages help to explain why so much discourse about African sacred groves concerns both moral disorder and ecological degradation – because the groves represent the ideal of an orderly and unified social-ecological system (Luning 2007).

Sacred Groves in the North Pare Mountains

The Pare range is part of the Eastern Arc Mountains, an archipelago of highlands that stretches from southeastern Kenya to southern Tanzania. These mountains are considered to be a biodiversity "hotspot" because of their relatively high proportion of endemic species (Myers et al. 2000). North Pare is an enormous tilted block with a ridgeline that looms high above the savanna (see Map 6.1). Most of the area's population lives on this mountain block. Its eastern side receives more than twice the rainfall of the western side because the mountains force the warm, moist air of the trade winds to rise, cool, and condense into rain. Reliable rainfall fostered the growth of forests with an eighty- to one-hundred-foot closed canopy in the North Pare highlands.

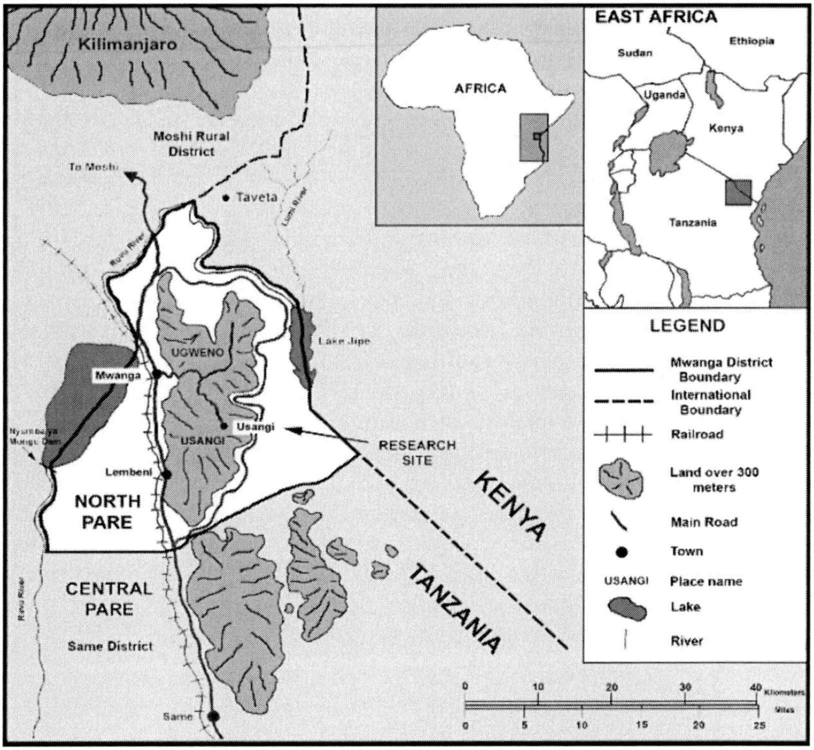

Map 6.1: The North Pare Mountains of Mwanga District, Tanzania

Indigenous trees grow only in the forest reserves on the highest peaks and in the hundreds of sacred groves that lie scattered throughout the densely settled zone between 4000 and 5300 feet above sea level. Although North Pare was once sparsely inhabited by foragers, significant environmental transformation began when farmers settled in the area between 250 and 790 CE. North Pare became an important node of the precolonial regional exchange network because its minerals gave it a comparative advantage over the volcanic areas to the north and west (Kimambo 1995). The clay and magnetite found in Pare is far superior to those found on the geologically young slopes of Kilimanjaro, so North Pare has long been the region's leading producer of pottery and iron (Iles et al. 2018). Precolonial agriculture and fuelwood consumption for these industries clearly transformed the vegetation cover of these mountains, and the evidence suggests a recurring pattern of forest clearance and secondary forest regrowth over 2000 years – not sacred groves as relics of an ancient primary forest (Heckmann 2014; Heckmann et al. 2014). Based on the general pattern of forest patches in the Eastern Arc mountains, however, biogeographers estimate that more than 76 percent of North Pare's preagricultural forest cover is now gone (Newmark 2002, 25, see also von Hellerman 2016 on the South Pare area).

According to oral histories, the major settlement was in Ugweno, at the northern end of North Pare. By collecting tribute in exchange for iron products, the ironworking Washana clan came to dominate the North Pare political economy by the thirteenth century (Kimambo 1969). Many Washana were killed in a late fifteenth-century coup by members of the subordinate Wasuya clan, who ambushed them outside of a sacred grove. As Kopytoff's model explains, when immigrants arrived in North Pare over the following centuries, men were assimilated into the Wasuya political system through initiation ceremonies held in sacred groves throughout the highlands. When the Wasangi clan arrived from Kenya in the seventeenth century, they adopted the Wasuya institutions and negotiated a partial autonomy from the Wasuya polity to form North Pare's southern settlement of Usangi. Sacred groves, and the political legitimacy they confer, have been both the glue and the solvent of politics in North Pare ever since.

Figure 6.1: Two Sacred Groves in Usangi, Tanzania: The Mshitu wa Sereni (large oval) and the Mpungi wa Semvua (smaller oval).

Figure 6.2: Skulls on Potsherds in a Sacred Grove, Usangi, Tanzania.
Note: The author did not enter this grove; it had been cut open by a road-building project.
The ancestors had not yet announced their desire to relocate at the time of the photo.

Sacred groves stand out both visually and culturally in the North Pare landscape as concentrations of both vegetation and meaning among its fields, houses, markets, roads, and schools (see Figure 6.1). According to the fundamental management rules, groves must be composed of only indigenous species and no vegetation may be cut (4). There are two kinds of sacred groves in Pare; a *mpungi* is a small shrine of trees, usually an acre or less in size, whereas a *mshitu* is a forest ranging from three to ten acres. Each mpungi contains the skulls of a particular lineage's ancestors, and there the living asks the dead for their blessings (see Figure 6.2). Young men are initiated into manhood in the mshitu forests, which are more powerful than the mpungi shrines—as well as far fewer in number—because they represent the ancestors of a ruling clan rather than an isolated lineage (5). The sanctions for violating a mshitu are therefore far more severe than those for a mpungi, and cursing someone by performing a sacrifice in a mshitu is commonly considered to be attempted murder. As one mshitu caretaker expressed its power to me, "if you go in without permission, I'll go give some beer and say my two words and you won't last two days!" More typically, however, elderly men sacrifice beer and meat to their ancestors to "cool" their anger and to forestall the disease, bad luck, drought, famine, and paranormal events that angry spirits are likely to send. The essential characteristic of a mshitu is its secrecy rather than its vegetation status, and it is for this reason that the idiom for "keeping secrets" and oath-taking in North Pare is *kula mshitu* — "eating a sacred forest (6)."

Each grove has a designated caretaker, typically the most senior male of a lineage or clan. These men are responsible for conducting sacrifices, organizing initiation rituals, and monitoring the integrity of their groves. Many of these elders also hold a variety of other local leadership roles. For example, in 1998 Hamisi Saidi of Chomvu village tended the Kwa Mbendi mpungi, allocated water from the irrigation furrow next to his grove, and served on the village Land Use Planning Committee. These were, he told me, aspects of his public role as an elder, not three separate institutional positions. This does not mean, however, that sacred grove caretakers form a coherent publicly recognized group with tightly bonded social capital. Their role depends on context, and people throughout North Pare invariably reckon who the caretaker should be for a given grove based on the logic of patrilineal kinship and the sort of action deemed necessary. Religious doctrine also shapes how these elders perform their caretaker duties. In my experience in North Pare, Lutheran and Evangelical Christians tend to neglect (or even destroy) their sacred groves, while Roman Catholics and Muslims maintain them. The Protestants see the groves as having negative cultural capital that conflicts with their religious identity, but the other groups see their groves as full of cultural capital that complements their religious practices. Actual ritual performance depends on the scale of the issues involved; as many caretakers expressed it, "Muslims go to the groves when they have a problem, and Christians go when they have a very big problem."

Environmental History of Sacred Groves in North Pare

According to the local authorities and the development agencies running projects in the 1990s, North Pare has suffered massive deforestation in recent decades driven by human population growth (*Guardian* 1997). This view is incorrect. North Pare's population grew rapidly over the twentieth century, and its farmers increased tree cover by planting exotic trees such as coffee and eucalyptus. The result is that the densities of both people and trees are now much higher than they were one hundred years ago (Gillson et al. 2003). Evidence for precolonial deforestation comes from the accounts of nineteenth-century travelers. One German explorer reported forests only in the nearly uninhabited northwestern quarter of North Pare and on the outer flanks of the massif. "Elsewhere," he wrote, "everything has been burned down for clearings, or, as on the higher zones of the mountains, the slopes are covered with low bushes, grass, or ferns" (Meyer 1891, 223). A missionary related that "there [were] few forests, and most of the hills and mountains are bare" in the area (Althaus 1894, 451). Finally, a geographer described the North Pare highlands as an agricultural mosaic: "The actual inhabited and cultivated area, which was originally well forested, and which still stands in scattered thick groups of trees, has long been developed [for agriculture]" (Baumann 1891, 200). These "groups of trees" are clearly the hundreds of sacred groves that punctuate the Pare landscape today, and which make up 8 percent of the land area of Mwanga District and 68 percent of its total forested area (Semgalawe 1996). Most of these (77 percent) lie in the densely settled zone between 4000 thousand and 4600 feet above sea level (Ylhäisi 2004, 117).

In my interviews in Usangi in 1997 and 2004, officials of the Tanzania Forestry Action Plan (a forestry program sponsored by the German government through the GTZ agency) often identified the sacred groves as remnants of North Pare's original forest. Indeed, the existence of patches of indigenous trees helped to justify the program. The project (active from 1991 to 2003) was based on the assumption that the "mountains used to be an area favourable to agriculture with large tracts of land under forest," but that recent population pressure had meant that "traditional environmental protection measures and farming practises could not ensure the sustainability of the natural resources" (GTZ 2003). This scenario compressed hundreds of years of landscape transformation into a crisis narrative demanding immediate action (Leach and Mearns 1996). The idea that North Pare's sacred groves were unchanged relics of a pristine forest appealed to both the development planners and the groves' caretakers, who still insist that their ancestors "carved" these groves from the larger forest as the founding acts of frontier settlement. Such statements about "carving" cannot be taken at face value, however. The identification of the groves as primeval forest may suit the goals of the different social actors attempting to restore a static harmony of culture and ecology, which they call (according to their preferred point of view) either "tradition" or "sustainability." But these groves owe their existence to tenurial legitimacy and political authority, and their creation narratives are tales about social relationships in flux rather than ecological descriptions.

It is likely that the mpungi exist today primarily because of the rapid segmentation of kin groups in precolonial Pare (Kimambo and Omari 1972), and also because of unintended ecological consequences. After arriving in the area, male elders established a sacrificial spot under a tree near their new homes (a place for cultural capital), which legitimized the immigrants' rights to land and their place in society (social capital). As the families grew and created new homesteads and farms, they were prone to segmentation. The sacrificial areas therefore became places that maintained lineage unity through collective rituals. This social process interacted with the dynamics of banana-based long-fallow agriculture (see Sheridan 2002) to produce sacred groves as coevolved features of the North Pare landscape. Many groves cover rocky outcrops or springs, and as farmers transformed the North Pare landscape with ax and fire, the groves grew from shrubs and trees left on nonarable areas. With a prohibition on cutting vegetation within a ritualized area, the margins of each sacred grove grew into thick walls of greenery rather quickly because of the availability of sunlight from multiple directions. Many of the trees commonly found in the sacred groves are fruit-bearing (see Table 1), so it is likely that many germinated from seeds carried there by birds and primates that still use them today as refuges and migration corridors. The result, as Baumann saw in 1890, is "scattered thick groups of trees" in a landscape of farms and pasture.

Species designation	**Vernacular name (in Chasu)**	**Fruit-bearing?**
Albizia gummifera versicolor	Mtanga	No
Albizia schimperana	Nyasutu	No
Bridelia micrantha	Mwira	Yes
Combretum schumannii	Muhama	Yes
Cordia abyssinica	Mringaringa	Yes
Croton macrostachyus	Mfirifiri	Yes, but poisonous to humans
Croton megalocarpus	Muhande	Yes
Diospyro mespiliformis	Mdulu	Yes
Dracaena fragrans (shrub)	Ithae	Yes

Ficus sycamorus	Mkuu	Yes
Ficus thonningii	Mvumo	Yes
Maerua triphylla	Mluhindi	Yes
Manilkara mochisia	Ihuu	Yes
Manilkara discolor	Mgambo	Yes
Syzigium guineense	Mlama wa Chasu	Yes

Table 1: Species Composition of the Sacred Groves of North Pare
Note: The species list is from Mzee Iddi Mkenga of Usangi. Species identification by Haruni Mbilinyi. Data on fruit edibility from Hines and Eckman (1993) and Plant Use Project database at https://uses.plantnet-project.org/.
Accessed December 9, 2022.

That these groves have, for the most part, persisted through a period of rapid population growth is testament to their importance in North Pare society. The people of North Pare are quite aware of the many ecosystem services that their mpungi and mshitu provide, and often point out the groves' ecological functions when discussing soil and water conservation projects (7). It is likely, however, that the groves contribute less to faunal biodiversity than they did a century ago. Vertebrate animals such as eagles and duikers figure prominently in North Pare oral histories, yet they are increasingly rare today. The predators and large herbivores were hunted out long ago, and the smaller species are fewer in number because the pasture and fallowed fields that once lay adjacent to the sacred groves have recently become intensively cropped fields. In 1982, only 18 percent of the groves in North Pare had been completely encircled by farms and houses, but 69 percent were circumscribed by 1997 (Ylhäisi 2004, 119). As is the case for other African forests, therefore, linking social history and ecology offers more insight than assuming that sacred groves are shards of a lost "balance of nature" (and culture).

Political History of Sacred Groves in North Pare

The sacred groves of North Pare are fine examples of ritual-ecological institutions on Africa's internal frontier, and North Pare's sacred groves have been engines for converting social, political, and cultural capital for at least five hundred years. While there was no central authority over sacrifices in the mpungi shrines, the mshitu forests served as basic political institutions throughout the region. By establishing a special council of male elders to supervise the content and practice of initiation, the chiefs of Ugweno directly controlled the means for youths to marry, bear children, achieve respect, and eventually have their skulls installed in the clan or lineage shrine. In the sixteenth century, the Wasuya chiefs of Ugweno lengthened the ceremony into a six-month period of intensive political, military, and cultural training backed by a series of oaths and ordeals (Kimambo 1969, 52). Clan histories assert that the mshitu forests were the primary means of social control and community-building because initiation ceremonies made immigrants socially legitimate. Usangi became an exception when the Wasangi clan secured the right to oversee their own initiation ceremonies, as long as they sent regular tribute to Ugweno. This was significant because the combination of ritual and political authority made the Wasangi the only clan in North Pare with the same sort of cultural capital as the Wasuya chiefs of Ugweno. Usangi broke away from Ugweno in the mid-eighteenth century with the help of mercenaries from the Mbaga chiefdom of South Pare.

The sacred groves of North Pare are central social institutions that link diverse cultural domains such as land tenure, moral order, and political legitimacy. Even today, each mpungi serves as a prominent reminder of a patrilineage's legitimate land ownership; the presence of skulls, for example, demonstrates how long a patrilineage has occupied its land. In local cultural terms, this institution is far more significant than the property rights delineated in statutory law. Another pragmatic function of the mpungi shrines is the maintenance of moral order. Parents regularly admonish their children to behave because their ancestors are watching them, and victims of injustice can seek revenge by sacrificing to the skulls. Many residents of Usangi say that sacrificing in a mpungi gives faster and more equitable justice than the local court does. For example, when someone stole a contact switch from a milling machine in January 1998, the miller led a goat down the main road through Usangi, loudly declaring that he was going to sacrifice in a mpungi the next day. The women in the marketplace began to shout, "Eee, this man is going to kill someone!" and the thief returned the part before nightfall.

While the smaller mpungi shrines mediate relationships with ancestors, the larger mshitu forests represent political and ecological relationships. The political functions of the mshitu made them critical resources for the ruling clans of the precolonial chiefdoms of North Pare, and not only because the initiation ceremonies formalized the chiefs' social control. In Usangi the legitimacy of the ruling Wasangi clan became increasingly precarious after they allowed their Wambaga allies from South Pare to settle on their border with Ugweno in the eighteenth century. One condition of the alliance was that the Wambaga had to return to their own mshitu forests in South Pare rather than participating in the Wasangi-dominated rituals in Usangi, but they could practice rainmaking and accept farmers' tribute for this service. The Wasangi also claimed that their own sacrifices brought rain; as one informant put it, "the old men went into the mshitu, and as soon as they came out it would begin to drizzle." The Wasangi failed to contain the cultural and political capital of their allies because rainmaking and tribute collection gave the Wambaga their own claim to political legitimacy, even without sacred groves (see Håkansson 1998; Sheridan in press). By the middle of the nineteenth century the Wambaga had formed a chiefdom that rivaled that of the Wasangi, and they were quickly outpacing the latter in wealth and influence. The Wambaga chief Kengia created new marketplaces to supply provisions, ivory, and slaves to caravans from the coast (von der Decken and Kersten 1978, 18). The guns, cloth, and wire that he received gave him a distinct advantage over his Wasangi rivals, and interchiefdom relations became increasingly fractious. Violence broke out during an initiation ceremony when the Wasangi assumed that Kengia's son was going to force his way into the mshitu forest at Mbale. A period of intense internecine warfare, lawlessness, and famine (now known as *Kibonda* throughout Pare) began in the early 1870s, culminating in 1891–92 with German intervention and the deaths of most of the Wambaga ruling family (8).

Colonial and Postcolonial Political Ecology

The politics of sacred groves in North Pare, and particularly their conversions of cultural capital into social and political capital, did not fade with the imposition of colonial rule; rather, their new roles reflected other processes of cultural and social change (9). Specifically, the history of Usangi's mshitu forests dovetailed with the rivalry between the Wambaga and Wasangi clans throughout the British colonial era. Horrified by exaggerated accounts of North Pare initiation rituals, the British prohibited the ceremonies throughout the Pare range in 1923 (Tanzania National Archives [hereafter TNA] 1733/23). Kin groups continued to sacrifice in their mpungi, and the Wasangi clan defended their mshitu as sacred areas, but this ban and the spread of Christianity and Islam undermined the political salience of the groves from 1923 to 1948. For most of this period the chief of Usangi Division was Sabuni Naguvu of the Wambaga clan. Since the establishment of indirect rule in 1926 had reduced the number of chiefs in Usangi Division from nine to three, the Wasangi clan was formally shut out of local government (TNA microfilm 11). The Wasangi had long been campaigning for recognition as the legitimate ruling clan of Usangi and they were outraged when the colonial administration allowed William Sabuni to succeed his father in 1948. As a result, they began to vigorously assert their control of the mshitu forests, the defining political symbol for the Wasangi, in repeated displays of what Spivak (2006) has called strategic essentialism—the strident assertion of a particular interpretation of tradition and authenticity (see also Spear 2003).

In 1951 William Sabuni convicted two Wasangi elders of supervising an illegal mshitu initiation and charged them the maximum fine of Tsh 200/= (£10) and six months of hard labor. Although the district commissioner waived the labor, the Wasangi were sufficiently incensed (and sophisticated) to appeal to an international authority, submitting a petition to the United Nations Trusteeship Council for a review of Sabuni's legitimacy. The major complaint in the petition was that the chief had "the intention of disallowing and killing this [initiation ceremony] as it is closely tied and associated with ruling affairs" (TNA 19/11/2B/196 and U.N. Trusteeship Council 1955). Before the U.N. had addressed the petition the colonial administration responded to the situation in 1952 by allowing the initiation ceremonies to take place in North Pare, but only under the direct supervision of the Wambaga chief (or his representative) (TNA 19/12/2A/II/233), who was to stand at the entrance to the mshitu forest and collect a fee from each initiate as he entered. This was a British attempt to foster the "tribal unity" that indirect rule demanded, but such direct control of a Wasangi ritual by a Wambaga chief was antithetical to the eighteenth-century political agreement between these two clans. While waiting for the United Nations to act, the Wasangi elders supervised illegal initiations, were often arrested, and repeatedly paid the maximum fine of Tsh 200/=. When the petition failed in 1955, the Wasangi leaders responded by recasting their campaign for control of the mshitu forests as an environmental issue.

The forests had long carried an indigenous political and ecological symbolism as bringers of rain, but colonial development programs contributed yet another layer of cultural capital. In the early 1950s Usangi was the focal point of the Pare Mass Literacy Scheme, which taught adults to read and write using soil and water conservation literature (Mason 1952; *Fundamentals of Adult Education* 1955–56a, 1955-56b). Specifically, district administrators wanted to protect water sources, ban livestock from steep slopes and hilltops, and mandate tree planting (TNA 517/A3/13/1). Adopting this rhetoric for their political campaign, the Wasangi accused Chief Sabuni's assistant, Sefu Mbwana, of grazing livestock and burning trees in the Mshitu wa Mbale forest. In 1931 Chief Sabuni had given Mbwana, then his personal bodyguard, permission to establish a cassava farm on the area of the forest that had been damaged by Lutheran missionaries (10). Mbwana planted *Eucalyptus globulus* and *Grevillia robusta* seedlings in 1937 after the district commissioner had ordered all Native Authority staff to plant trees on the bare summits of North Pare. But from 1937 to 1956 Mbwana repeatedly cut and burned his trees to make charcoal on the spot and allowed the stumps to coppice (Mbwana n.d.). In July 1956 the Wasangi, by then organized into a political party called the Wasangi Union, wrote to the Pare Council formally accusing Mbwana of destroying a sacred grove (TNA 517/A2/2/58). The Council ordered Mbwana to stop cutting trees, and the Wasangi brought powerful allies to survey the damage to their grove. Three chiefs from South Pare visited Usangi and sent a sympathetic report to the Pare Council stressing that the area "is of great value, and must be preserved as it was by our fathers, because it gives water, shade, and clean air, and it preserves the hilltop of Mbale" (TNA 517/A2/2/88). The Wasangi won the legal skirmish, and in 1960 they installed Sangiwa Mtengeti II as chief of Usangi and revived initiation ceremonies in all five of Usangi's mshitu forests. These initiations are still held every year in September and usually involve about thirty initiates in a secret six-day training program.

After the transformation of local government from chiefdoms to a local administrative structure under direct government control, the mshitu forests lost their role as capital conversion mechanisms. Throughout Usangi people still refer to the mshitu as part of the ongoing discussion about who would be the chief of Usangi today if Nyerere had not changed the structure of local government (see Sheridan 2004), and both the mshitu and mpungi groves have maintained their importance as areas for initiation and sacrifice. However, after independence they no longer benefited from political protection, and many informants reported that in the early 1960s farmers were quick to deduce that colonial era conservation laws no longer applied. Water sources, swamps, riverbanks, and sacred groves were invaded and cultivated throughout North Pare. Many groves, such as the Mshitu wa Kena, lost up to half of their area to encroachment by farmers hungry for their rich soil. Some mpungi were completely destroyed.

Much of this decline in the value of the groves was based on (and justified by) religious differences. By the 1960s most people in Usangi identified themselves as either Christian or Muslim, and attitudes toward the sacred groves followed this cultural and generational divide. In general, most of the men who now value sacred groves are Muslims. Most of the groves that have been cut or destroyed were those managed by Christians, and farmers, in fact, often defend grove encroachment as a demonstration of Christian fervor. At public meetings in Usangi I often heard many Christians and young men say that the sacred forests were "absurdities," and that their elderly caretakers should "go with the times." The old men's ironic rejoinder was usually some version of the question, "if we go with the times, where will that take us?"

The caretakers' struggle to maintain their groves has become increasingly difficult since independence. Although the local government regulations explicitly protect the sacred groves, the legal procedure to stop someone from cutting a grove is cumbersome, time-consuming, and expensive in terms of both cash and political capital. The process became particularly difficult in the 1970s, when the Tanzanian government urged farmers to maximize production. Under such circumstances, the caretakers say, "the farmer could not be wrong," and the offenders just had to insist that they had been following the government policy that "politics is agriculture" (TANU 1972). The caretakers' defense of the groves became even more precarious in the late 1980s, when the Tanzanian government began to liberalize its economy under pressure from the World Bank and International Monetary Fund. The value of the shilling plummeted, and the fines that offenders paid in court did not keep up with inflation. In the 1980s and 1990s, those found guilty of cutting a tree in a sacred grove paid a fine of Tsh 1000/=. This was a heavy penalty (about US$100) when the district's Protected Areas By-laws were passed in 1984, but by 1998 was worth only USD$0.75. And given the fact that the plaintiff must supply the court with paper, pens, and the occasional bribe, sacred grove caretakers find legal proceedings expensive, while offenders find them cheap compared to the profits to be gained from a good-sized log. If the judgment goes against the plaintiff, a caretaker often must reimburse the defendant's legal costs, so caretakers must carefully evaluate the chances of winning and the costs of losing.

The latest major development for Usangi's sacred groves involved the German-funded Tanzania Forestry Action Plan–North Pare (TFAP), which from 1992 until 2003 promoted terracing, agroforestry, and the use of fuel-efficient stoves through village-based land use planning committees. TFAP administrators soon realized that the project should involve the area's major examples of indigenous environmental conservation, the sacred groves. Tanzania's forestry policy was also calling for their protection (United

Republic of Tanzania 1997, 11) (11). After a series of discussions with sacred grove caretakers, German and local TFAP administrators organized a meeting of more than two hundred elderly men from all over North Pare in May 1996 to discuss the ecological and cultural values of the groves. Participants agreed to conserve the groves because they house the skulls of the ancestors, bring rain, and protect the aquifers of many springs and water sources. The meeting participants decided that the best ways to protect the mpungi and mshitu were by formalizing the customary fines for violating a sacred grove, marking their borders with *Dracaena* plants, and "strengthening" them by planting indigenous trees (Semvua 1996; TFAP 1997). Although the customary fines still have no legal basis as district by-laws (which require ratification by the central government, a notoriously slow process), this meeting and the dissemination of the schedule of fines have given caretakers a powerful tool to punish violators. As the grove caretakers themselves are quick to point out, however, the fines may not withstand close legal scrutiny.

On the whole, this amalgam of rational bureaucratic process and ritualized forest management has been successful. Sacred grove caretakers told me in 2004 that the formalization of sanctions had strengthened their management capacity and prevented grove encroachment. A 2003 ecological assessment of the sacred groves in Ugweno (the northern half of the North Pare massif) suggested that the new rules and adequate rainfall had hastened their ecological recovery since the mid-1990s (Ylhäisi 2004, 121). As is often the case, however, this new institutional dynamic has had some unexpected consequences, caused in this instance by the tenurial implications of tree planting and, in response, the custodial behavior of some local caretakers. As in many parts of Africa, the act of planting trees in Pare creates long-term use-rights, yet kin groups claim ownership (Fortmann 1988). The land tenure system of North Pare features seasonal land loans, landholding patrilineages, and the omnipresent potential for the state to assert its ultimate ownership of land. Farmers who borrow land, and particularly female farmers, are enjoined from planting trees because this can create use-rights that last decades. Although they sought to "strengthen" the groves, TFAP and the local government did not recognize that these tenurial muddles apply to sacred groves.

There was no source for indigenous tree seedlings in North Pare, so TFAP provided ficus cuttings—which propagate vegetatively—for the groves' caretakers to "strengthen" the groves under the supervision of the Divisional Forestry Officer and TFAP extensionists in 1998. Most of these cuttings failed to take root and dried up, however, and many of those that did sprout roots were found later with dry roots in the air and their leaves underground. TFAP administrators thought that encroaching farmers were preventing the cuttings from flourishing by inverting them. But when the caretakers were asked about this issue, they explained that they had sabotaged the project themselves. In effect, the central government's monopoly on legitimate economic capital trumped local institutions for social and cultural capital. Sacred grove caretakers are keenly aware of Tanzanian land tenure ambiguities, which they usually summarize as, "I own what I have planted, but the land belongs to the government." They point to their government's authoritarian style, the nationalization of private property after the 1967 Arusha Declaration, and the many examples of uncompensated land seizures. They also apply the logic of the tree tenure system to their own situation when they evaluate their rights to manage sacred groves. When TFAP asked them to plant trees, the caretakers complied, but reasoned (correctly) that forest administration would revert to government officials after the German project left. TFAP's use-rights to those ficus saplings would therefore pass to the local government, which meant that each "strengthened" sacred grove would be encircled by government ownership. It was not too difficult to imagine a future government order declaring the sacred groves to be Forest Reserves, and no longer kin group property.

The key issue is that the sacred groves of North Pare connote social order and morality – and the TFAP intervention upset the logic and meaning of local control. As one elder said,

> Why does TFAP need to know the number of trees and the size of a mpungi? Are they going to sue somebody if it changes? The elders think they want to take them from us . . . We don't have development now, we only have fear and doubt.

The bureaucratic imperative to rationalize sacred grove management foundered in a tenurial morass that the administrators did not understand or anticipate. Efforts to reinforce existing management patterns, however, were more welcomed and successful.

Conclusions

This account has shown how the ecological status of African sacred groves is linked not only to their spiritual significance, but also to politics, economics, and legal processes. As is often the case in sub-Saharan Africa, land tenure in Pare is a negotiated social process rather than a structure of rules and rights. Because various social relationships of land use, kinship, political authority, and metaphysics intersect in African sacred groves, the trees have been particularly vulnerable to the institutional changes and capital conversion strategies introduced by colonial and postcolonial states. The sacred groves of North Pare were once a means to institutionalize kin group rituals and mechanisms for social control in polities that were expanding by assimilating immigrants. The groves symbolized the economic, social, political, and cultural capitals of legitimate land tenure, political authority, and moral order. The social and cultural transformations of colonial rule and independence did not replace these functions of the groves; rather, they added new layers of meaning. The groves' latest layer of meaning is their value as examples of indigenous conservation, in the pursuit of which their caretakers have been able to negotiate complex relationships with the local government and an internationally funded development agency. These layers of meaning contribute to decisions to cut or preserve trees, so that social history and political ecology have tangible effects on the ecological dynamics of culturally significant vegetation patches.

The complexities of local history and politics surrounding the sacred groves should make us cautious, but moderately optimistic, about their long-term survival in North Pare and other parts of Africa. As governments and conservation agencies struggle to create community-based conservation that protects biodiversity and promotes sustainable livelihoods, sacred groves are, in many ways, ideal for such efforts. It must be remembered, however, that the cultural values and institutions that maintain sacred groves are perhaps analogous to, but not really the equivalent of, the scientific concept of an ecosystem (Berkes et al. 1998). Any link between sacred groves and conservation that implies either an ecological or social equilibrium is probably misguided, since much of the data on sacred grove ecology, social organization, and meaning in Africa show contestation, dynamism, and transformation rather than stasis. Tropical forest patches often exist in states of ecological flux, African communities are not characterized by uniform values and organization, and the meanings of sacred groves are often multiple, overlapping, and ambiguous. Applying Bourdieu's approach to capital conversion does admittedly impose a perspective from Western neo-classical economics, especially its assumption that self-interested individual strategies form a larger system, which then distorts indigenous meanings and values. This chapter has attempted to demonstrate that institutional analysis plus ethnographic contextualization can overcome this flaw and describe African sacred groves more accurately than the "relic theory."

Efforts to decentralize and democratize resource management in Africa are often fraught with contradictions such as these (Manor 2005). Bureaucratic demands for communities with marked boundaries and permanent membership can foster divisive politics in which once-flexible social networks harden into blocs of "stakeholders" and "outsiders" (Geschiere 2004). As this case study of North Pare shows, the political and historical dynamics of African sacred groves can make their incorporation into conservation policy problematic at best. The complex and overlapping ecological, sociopolitical, and moral aspects of sacred groves are not reducible to bureaucratic planning, and rationalizing these sites in order to enhance biodiversity can threaten the very local land management institutions that conservationists want to support (12). Sacredness, as Roy Rappaport (1999) emphasized, is usually unquestionable, dangerous, and secret. This does not make for a good fit with the transparency, accountability, and empiricism that most conservation bureaucracies require. The potential for the co-management of African sacred groves is now emerging as a result of policy reforms that are decentralizing land and tree tenure to local authorities (Alden Wily 2008, 2018a, 2018b; Blomley and Ramadhani 2007), but few conservation institutions have engaged African sacred sites on their own terms. This would require administrators to investigate and document the complex (and often disputed) history, meanings, and legal status of African sacred groves. What is needed are policies based on both social and natural science, a hybrid sort of conservation in which sacred grove caretakers would need to be empowered as equal stakeholders alongside ecologists and bureaucrats (13). Finally, it would be necessary for conservation practitioners to accept (and perhaps embrace?) a processual approach that includes negotiated social dynamics and shifting meanings rather than steady states of ecology and society.

Tanzanian government agencies and NGOs could, for example, create such a hybridized co-management system in North Pare by agreeing to caretakers' requests that outsiders offer sacrifices (e.g., of beer or meat) in sacred groves when introducing new management procedures and goals. This would be a relatively cheap way for agencies to gain substantial cultural capital and local legitimacy. Development agencies could help kin groups undertake the costly process of having their groves surveyed in order to get title deeds. Tree nurseries could stock indigenous species, and the status of *Dracaena* as a boundary marker could be formalized in the area's by-laws (see Sheridan 2023). Outside agencies could facilitate a forum for the area's religious leaders to discuss how religion relates to environmental protection, and this could enhance the social capital and trust among these occasional rivals. Such discourses and practices would have the potential of redefining conservation in North Pare, where most farmers often say that the Kiswahili term for conservation, *hifadhi*, really means government ownership and mismanagement (14).

The particular local history of ecology, society, and culture means that there is unlikely to be a single blueprint for managing sacred groves in Africa. The vital issue facing scholars, conservationists, and sacred grove caretakers is not how conservation policy can absorb the beneficial ecological and social functions of sacred groves in Africa; rather, the challenge is to allow the diversity and dynamics of African sacred groves to transform the nature of conservation.

Acknowledgments

Fieldwork for this chapter was supported by the Wenner-Gren Foundation for Anthropological Research and the Fulbright program in 1997–98, and by the University of Vermont in 2004. It would not have been possible without the assistance of Haruni Mbilinyi, Ashura Ayubu, and Fatu Ramadhani, and the inspiration of Mzee Hasani Sefu Changaja. The manuscript was greatly improved by the editing skill of Ella Kusnetz. The original version of this substantially revised chapter was published as a 2009 article in *African Studies Review* 52(1): 73-98. Reprinted with permission.

Notes

[1] The kaya forests of coastal Kenya are particularly well-documented sacred groves. See Nyamweru and Kimaru (2008), Parkin (1991), Spear (1978), and Willis (1996).
[2] For elements of this narrative, see Anoliefo et al. (2003), Bagine (1998), Chidhakwa (2003), Mwihomeke et al. (1998), Okeke (1999), Orlowska and Klepeis (2018), Sibanda (1997), Tchamie (2000), Warren and Pinkston (1998), and Ylhäisi (2000).
[3] Authors such as Gadgil and Guha (1992, 38), Gokhale et al. (1998, 366), and Hughes (1994, 170) emphasize that sacred groves correspond with horticultural modes of production worldwide.
[4] The rule against cutting trees in a sacred grove is usually interpreted quite strictly by their caretakers. Even if a falling tree hits a house, the homeowner cannot clear away the tree without the caretaker's permission and the necessary sacrifices.
[5] This chapter does not describe initiation rituals because their content has no direct bearing on the political ecology of North Pare. The ceremonies consisted of a series of physical hardships and lessons about law, politics, and sexuality (Lebulu 1979).
[6] Oath-taking in Pare typically involves ingesting an herbal concoction that will, in theory, cause one's stomach to burst if the oath is violated.

[7] Much of this discourse emphasizes that the ancestors had deep ecological wisdom when they established the groves, and that current conservation efforts should reinforce rather than replace indigenous methods. In the conservation meetings I attended in 1997–98, these subtle critiques were consistently ignored by development planners.

[8] This was part of a widespread regional process of ecological, economic, and political disruption throughout East Africa (Kjekshus 1977).

[9] For a review of the colonial political economy of Pare, see Kimambo (1991).

[10] Labor and taxation policies were major agents of change in the German colonial period (ca. 1885–1917), but in the North Pare highlands missionaries were more influential. Hans Fuchs established a Leipzig Lutheran Mission station in July 1900 and developed a reputation for cutting down sacred groves during his long career in North Pare (1900–1934). Because North Pare was largely deforested by the early twentieth century, he built churches and houses with local sacred logs and secular timber from Usambara and Kilimanjaro (Fuchs 1901, 38).

[11] The policy calls for village governments to be responsible for forests not already under central government control as Forest Reserves. The policy states that "existing traditional forests with established indigenous management systems will be protected from any disturbance and new traditional forests will be encouraged to be established."

[12] For examples of calls to rationalize African sacred groves, see Dorm-Adzobu et al. (1991), Mgumia and Oba (2003), and Okafar and Ladipo (1995).

[13] The call for hybrid scientific research can be found in Batterbury, Forsyth, and Thomson (1997).

[14] Undoubtedly such efforts would also foster new complexities in regard to neo-traditionalism (Spear 2003) and the simplification of belief systems (Ranger 1999). For examples of efforts to create ostensibly "traditional" organizations for planting trees and maintaining forest cover in Africa by reference to notions of "sacredness," see Baker (1931) and Daneel (2001). These case studies (from colonial Kenya and independent Zimbabwe, respectively) showed how difficult it is for outsider conservationists to synthesize disparate value systems and principles of social organization. Both efforts created supposedly "traditional" secret brotherhoods of men, and both had short-term success and long-term failure because of symbolic and organizational contradictions.

Archival Materials

Mbwana, S. n.d. Personal diary, 1924–1965. Notebook in possession of Mzee Godwin Ndama, Kiriche, Usangi, Tanzania. Copy in possession of the author.

Semgalawe, S. 1996. "Baseline Survey of Mwanga District, Part 1: Review of Existing Information." Mwanga: NRBZ Development Program.

Tanzania National Archives (TNA), Dar es Salaam

TNA 19/12/2A/II/233. "Sheria za Ngoma ya Wenyeji wa Upare." 1952.

TNA 19/11/2B/196. Wasangi petition to the UNO delegation, September 11, 1951.

TNA 517/A2/2/58. Letter, President of the Wasangi Union to Chairman, Pare Council, July 28, 1956.

TNA 517/A2/2/88. "Report ya Msitu Mbale," Usangi, September 13, 1956.

TNA Microfilm 11, Pare District Book. "Notes on the structure of tribal government in Upare." 1934.

TNA 517/A3/13/1. Main Points of Smithyman's 5 Years [*sic*] Plan 1953–1957.

TNA 1733/23. Moshi District Annual Report. 1923.

TFAP-North Pare archives, Mwanga, Tanzania

Semvua, A. F. 1996. "Mishitu and Mpungi: Exposure Draft." TFAP–North Pare files, GTZ Mwanga.

TFAP - North Pare. 1997. "Mpango wa ulinzi, utunzaji, na uendelezaji mpungi na mishitu, Tarafa ya Usangi na Ugweno." [Plan for protection, management, and expansion of mpungi and mshitu, Usangi and Ugweno Divisions].

References

Adams, William, and David Hulme. 2001. "Conservation and Community: Changing Narratives, Policies and Practices in African Conservation." In: *African Wildlife and Livelihoods*. Edited by David Hulme and Marshall Murphree, 9–23. Cape Town: David Philip.

Alden Wily, Liz. 2008. "Are Sacred Forests Safe? An Examination of the Legal Status of Forests in sub-Saharan Africa." In: *African Sacred Groves: Ecological Dynamics and Social Change*. Edited by Michael Sheridan and Celia Nyamweru, 207–220. Oxford: James Currey.

—. 2018a. "Collective Land Ownership in the 21st Century: Overview of Global Trends." *Land* 7 (2): 68.

—. 2018b. "Customary Tenure: Remaking Property for the 21st Century." In: *Comparative Property Law*. Edited by Michele Graziadei and Lionel Smith, 458-477. Cheltenham, UK: Edward Elgar Publishing.

Althaus, Georg. 1894. "Untersuchungsreise nach dem Ugueno-Gebirge." *Evangelisch-Lutherisches Missionsblatt* (24): 449–53.

Anane, Mike. 2003. "Religion and Conservation in Ghana." In: *Implementing Agenda 21.* Edited by Leyla Alvanak and Adrianne Cruz. pp. 99-108. New York: United Nations Non-Liaison Services.

Anoliefo, G. et al. 2003. "Environmental Implications of the Erosion of Cultural Taboo Practices in Awka-South Local Government Area of

Anambra State, Nigeria." *Journal of Agricultural and Environmental Ethics* 16: 281–96.

Awuah-Nyamekye, Samuel. 2009. "Salvaging Nature: The Akan Religio-Cultural Perspective." *Worldviews: Global Religions, Culture, and Ecology* 13(3): 251-82.

Bagine, R. 1998. "Biodiversity in Ramogi Hill, Kenya, and its Evolutionary Significance." *African Journal of Ecology* 36(3): 251–63.

Baker, Richard. 1931. *Men of the Trees*. New York: Dial Press.

Batterbury, Simon. et al. 1997. "Environmental Transformations in Developing Countries: Hybrid Research and Democratic Policy." *Geographical Journal* 163(2): 126–32.

Baum, Robert. 1999. *Shrines of the Slave Trade*. New York: Oxford University Press.

Baumann, Oscar. 1891. *Usambara und seine Nachbargebiete*. Berlin: Dietrich Reimer.

Berkes, Fikret, et al. 1998. "Exploring the Basic Ecological Unit: Ecosystem-like Concepts in Traditional Societies." *Ecosystems* 1: 409–15.

Blomley, Tom, and Hadija Ramadhani. 2007. "Participatory Forest Management in Tanzania – an Overview of Status, Progress, and Challenges Ahead." *The Arc Journal* 21: 3-5.

Bourdieu, Pierre. 1986. "The Forms of Capital." In: *The Handbook of Theory and Research for the Sociology of Education*. Edited by J. Richardson, 241-258. New York: Greenwood Press.

Bravman, William. 1998. *Making Ethnic Ways*. Portsmouth, N.H.: Heinneman.

Byers, Bruce. et al. 2001. "Linking the Conservation of Culture and Nature: A Case Study of Sacred Forests in Zimbabwe." *Human Ecology* 29(2): 187–218.

Campbell, Michael O'Neal. 2004. "Traditional Forest Protection and Woodlots in the Coastal Savannah of Ghana." *Environmental Conservation* 31(3): 225–32.

Cardelús, Catherine. et al. 2019. "Human Disturbance Impacts the Integrity of Sacred Church Forests, Ethiopia." *PLOS One* 14(3): e0212430.

Castro, Peter. 1990. "Sacred Groves and Social Change in Kirinyaga, Kenya." In: *Social Change and Applied Anthropology*. Edited by Miriam Chaiken and Anne Fleuret, 277–289. Boulder : Westview Press.

Chevalier, A. 1933. "Les Bois Sacrés des Noirs, Sanctuaries de la Nature." *Compte Rendue de la Société de Biogeographie* 1933: 37.

Chidhakwa, Zvidzai. 2003. "Traditional Institutions Manage their Nyakwaa and Chizire Forests in Chimanimani, Zimbabwe." *Policy Matters* 12: 132–40.

Chouin, Gérard. 2002. "Sacred Groves in History." *IDS Bulletin* 33(2): 39–46.

Colding, Johan, and Carl Folke. 2001. "Social Taboos: 'Invisible' Systems of Local Resource Management and Biological Conservation." *Ecological Applications* 11(2): 584–600.

Colson, Elizabeth. 1997. "Places of Power and Shrines of the Land." *Paideuma: Mitteilungen zur Kulturkunde* 43: 47–57.

Comaroff, John, and Jean Comaroff. 2009. *Ethnicity, Inc.* Chicago: University of Chicago Press.

Cormier-Salem, Marie-Christine, and Thomas J. Bassett. 2007. "Introduction: Nature as Local Heritage in Africa: Longstanding Concerns, New Challenges." *Africa* 77(1): 1-17.

Daneel, Marthinus. 2001. *African Earthkeepers*. Maryknoll, N.Y.: Orbis Books.

Decher, Jan. 1997. "Conservation, Small Mammals, and the Future of Sacred Groves in West Africa." *Biodiversity and Conservation* 6: 1007–1026.

Dorm-Adzobu, Clement, et al. 1991. *Religious Beliefs and Environmental Protection: The Malshegu Sacred Grove in Northern Ghana*. Nairobi: ACTS Press and World Resources Institute.

Dudley, Nigel. et al. 2005. *Beyond Belief: Linking Faiths and Protected Areas to Support Biodiversity Conservation*. Gland, Switzerland: World Wide Fund for Nature.

Elliott, Jennifer, and Michael O'Neal Campbell. 2002. "The Environmental Imprints and Complexes of Social Dynamics in Rural Africa: Cases from Zimbabwe and Ghana." *Geoforum* 33: 221–37.

Ellis, Stephen, and Gerrie ter Haar. 2004. *Worlds of Power: Religious Thought and Political Practice in Africa*. Oxford: Oxford University Press.

Fairhead, James, and Melissa Leach. 1996. *Misreading the African Landscape*. Cambridge: Cambridge University Press.

—. 1997. "Culturing Trees: Socialized Knowledge in the Political Ecology of Kissia and Kuranko Forest Islands of Guinea." In: *Nature is Culture*. Edited by Klaus Seeland, 7-18. London: Intermediate Technology Development Group Publishing.

Fortes, Meyer. 1945. *The Dynamics of Clanship among the Tallensi*. London: Oxford University Press.

Fortmann, Louise. 1988. "The Tree Tenure Factor in Agroforestry with Particular Reference to Africa." In: *Whose Trees? Proprietary Dimensions of Forestry*. Edited by John Bruce and Louise Fortmann, 16–33. Boulder, Colo.: Westview Press.

Frazer, James. 1890. *The Golden Bough*. London: MacMillan and Company.

Fuchs, Hans. 1901. "Grundung der neuen Station Schigatini in Nord-Pare." *Evangeslisches-Lutherisches Missionsblatt* 1901: 38–43.

Fundamentals of Adult Education. 1955–56a. "UNESCO Associated Projects–X: Community Development in the Pare District of Tanganyika, 1950–54." *Fundamentals of Adult Education* 7: 160–67.

—. 1955–56b. "UNESCO Associated Projects–X: Community Development in the Pare District of Tanganyika, 1950–54." *Fundamentals of Adult Education* 8: 31–35.

Gadgil, Madhav and Ramachandra Guha. 1992. *This Fissured Land*. Berkeley: University of California Press.

Galiani, Sebastian, and Itai Sened. 2014. *Institutions, Property Rights, and Economic Growth: The Legacy of Douglass North*. New York: Cambridge University Press.

Gerdén, C., and S. Mtallo. 1990. *Traditional Forest Reserves in Babati District, Tanzania*. International Rural Development Centre, Working Paper No. 128. Uppsala: Swedish University of Agricultural Sciences.

Geschiere, Peter. 2004. "Ecology, Belonging and Xenophobia: The 1994 Forest Law in Cameroon and the Issue of 'Community.'" In: *Rights and the Politics of Recognition in Africa*. Edited by Harri Englund and Francis Nyamnjoh, 237–259. London: Zed Books.

Gillson, Lindsey, Michael Sheridan, and Dan Brockington. 2003. "Representing Environments in Flux: Case Studies from East Africa." *Area* 35(4): 371–89.

Gokhale, Y. et al. 1998. "Sacred Woods, Grasslands, and Waterbodies as Self-Organized Systems of Conservation." In: *Conserving the Sacred for Biodiversity Management*. Edited by P. Ramakrishnan, K. Saxena, and U. Chandrashekara, 365–396. Enfield, N.H.: Science Publishers.

Greene, Sandra. 2002. *Sacred Sites and the Colonial Encounter*. Bloomington: Indiana University Press.

GTZ. 2003. *Tanzania Forestry Action Plan North Pare Project*. www.gtz.de/themen/projekt.

The *Guardian*. 1997. "Population Explosion Cause of Soil Degradation in Pare, Dar es Salaam." October 24.

Guilllot, Bernard. 1980. "La Creation et la Destruction des Bosquets Koukouya, Symboles d'une Civilisation et de son Declin." *Cahiers O.R.S.T.O.M. Serie Sciences Humaines* 17 : 177–189.
Håkansson, N. Thomas. 1998. "Rulers and Rainmakers in Precolonial South Pare, Tanzania : Exchange and Ritual Experts in Political Centralization." *Ethnology* 37(3): 263–83.
Heckenberger, Michael, et al. 2003. "Amazonia 1492: Pristine Forest or Cultural Parkland?" *Science* 301: 1710–14.
Heckmann, Matthias. 2014. "Farmers, Smelters and Caravans: Two Thousand Years of Land Use and Soil Erosion in North Pare, NE Tanzania." *CATENA* 113: 187-201.
Heckmann, Matthias;. et al. 2014. "Human-Environment Interactions in an Agricultural Landscape: A 1400-Yr Sediment and Pollen Record from North Pare, NE Tanzania." *Palaeogeography, Palaeoclimatology, Palaeoecology* 406: 49-61.
Hines, Deborah and Karlyn Eckman. 1993. *Indigenous Multipurpose Trees of Tanzania*. Ottawa: Cultural Survival Canada, Ottawa. www.fao.org.
Hughes, J.D. 1994. *Pan's Travail: Environmental Problems of the Ancient Greeks and Romans.* Baltimore: Johns Hopkins University Press.
Iles, Louise;. et al. 2018. "Iron Production in North Pare, Tanzania: Archaeometallurgical and Geoarchaeological Perspectives on Landscape Change." *African Archaeological Review* 35: 507-530.
Juhe-Beaulaton, Dominique. 2010. *Forets Sacrees Et Sanctuaires Boises: Des Creations Culturelles Et Biologiques (Burkina Faso, Togo, Benin).* Paris: Karthala.
Juhé-Beaulaton, Dominique, and Matthieu Salpeteur. 2018. "Sacred Groves in African Contexts (Benin, Cameroon): Insights from History and Anthropology." In: *A History of Groves*. Edited by Jan Wouldstra, and Colin Roth, 133-148. London: Routledge.
Kimambo, Isaria. 1969. *A Political History of the Pare of Tanzania, c. 1500–1900*. Nairobi: East African Publishing House.
—. 1991. *Penetration and Protest in Tanzania: The Impact of the World Economy on the Pare, 1860–1960*. London: James Currey.
—. 1995. "Environmental Control and Hunger in the Mountains and Plains of Northeastern Tanzania." In: *Custodians of the Land* Edited by Greg Maddox, James Giblin, and Isaria Kimambo, 71–95. Athens: Ohio University Press.
Kimambo, Isaria, and Cuthbert Omari. 1972. "The Development of Religious Thought and Centres among the Pare." In: *The Historical Study of African Religion*. Edited by Terrence Ranger and Isaria Kimambo, 111–121. Berkeley: University of California Press.

Kjekshus, Helge. 1977. *Ecology Control and Economic Development in East African History*. Berkeley: University of California Press.

Klepeis, Peter. et al. 2016. "Ethiopian Church Forests: A Hybrid Model of Protection." *Human Ecology* 44(6): 715-730.

Kopytoff, Igor. 1988. "The Internal African Frontier: The Making of African Political Culture." In: *The African Frontier*. Edited by Igor Kopytoff, 3–85. Bloomington: Indiana University Press.

Larson, Anne, and Jesse Ribot. 2005. "Democratic Decentralisation through a Natural Resource Lens: An Introduction." In: *Democratic Decentralisation through a Natural Resource Lens*. Edited by Jesse Ribot and Anne Larson, 1–25. London: Routledge.

Leach, Melissa and Robin Mearns. Eds. 1996. *The Lie of the Land*. London: International African Institute.

Lebbie, Aiah and Mark Schoonmaker Freudenberger. 1996. "Sacred Groves in Africa: Forest Patches in Transition." In: *Forest Patches in Tropical Landscapes*. Edited by John Scholhas and Russell Greenberg, 300–323. Washington, D.C.: Island Press.

Lebbie, Aiah and Raymond Guries. 2008. "The Role of Sacred Groves in Biodiversity Conservation in Sierra Leone." In: *African Sacred Groves: Ecological Dynamics and Social Change*. Edited by Michael Sheridan and Celia Nyamweru, 42-61. Oxford: James Currey.

Lebulu, Josaphat. 1979. "Religion as the Dominant Element of the Superstructure among the Pare of Tanzania." *Social Compass* 36: 417–59.

Luning, Sabine. 2007. "Ritual Territories as Local Heritage? Discourse on Disruptions in Society and Nature in Maane, Burkina Faso." *Africa* 77(1): 86-103.

Madeweya, K., et al. 2004. "Sustainable Management of Sacred Forests and their Potential for Eco-tourism in Zanzibar." *Shinrin Sogo Kenkyujo kenkyu hokoku [Bulletin of the Forestry and Forest Products Research Institute]* 3(1): 33–48.

Manor, James. 2005. "User Committees: A Potentially Damaging Second Wave of Decentralisation?" In: *Democratic Decentralisation through a Natural Resource Lens*. Edited by Jesse Ribot and Anne Larson. London: Routledge. 192–231.

Martin, Adrian, Anne Akol, and Nicole Gross-Camp. 2015. "Towards an Explicit Justice Framing of the Social Impacts of Conservation." *Conservation and Society* 13(2): 166-178.

Mason, H. 1952. "Progress in Pare." *Corona* 4(6): 212–19.

Mbiti, John. 2015. [Second Edition]. *Introduction to African Religion*. Long Grove, IL: Waveland Press.

Meyer, Hans. 1891. *Across East African Glaciers*. London: George Philip and Son.
Mgumia, F. and G. Oba. 2003. "Potential Role of Sacred Groves in Biodiversity Conservation in Tanzania." *Environmental Conservation* 30(3): 259–65.
Mwihomeke, Steven et al. 1998. "Traditionally Protected Forests and Nature Conservation in the North Pare Mountains and Handeni District, Tanzania." *Journal of East African Natural History* 87: 279–90.
Myers, Norman et al. 2000. "Biodiversity Hotspots for Conservation Priorities." *Nature* 403: 853–858.
Netting, Robert. 1969. "Marital Relations in the Jos Plateau of Nigeria." *American Anthropologist* 71: 1037–1045.
Newmark, William. 2002. *Conserving Biodiversity in East African Forests*. Berlin: Springer Verlag.
North, Douglass. 1990. *Institutions, Institutional Change and Economic Performance*. Cambridge: Cambridge University Press.
Nyamweru, Celia. 1996. "Sacred Groves Threatened by Development: The Kaya Forests of Kenya." *Cultural Survival Quarterly* 20(3): 19–21.
—. 2012. "Natural Cultural Sites of Kenya: Changing Contexts, Changing Meanings." *Journal of Eastern African Studies* 6(2): 270-302.
Nyamweru, Celia et al. 2008. "The Kaya Forests of Coastal Kenya: 'Remnant Patches' or Dynamic Entities?" In: *African Sacred Groves: Ecological Dynamics and Social Change*. Edited by Michael Sheridan and Celia Nyamweru. Oxford: James Currey. 62–86.
Nyamweru, Celia and Elias Kimaru. 2008. "The Contribution of Ecotourism to the Conservation of Natural Sacred Sites: A Case Study from Coastal Kenya." *Journal for the Study of Religion Nature, and Culture* 2: 327–50.
Okafar, J.C. and D.O. Ladipo. 1995. "Fetish Groves in the Conservation of Threatened Flora in Southern Nigeria." In *Conservation of Biodiversity in Africa*. Edited by L.A. Bennun, R.A. Aman, and S.A. Craftes, 167–179. Nairobi: National Museums of Kenya.
Okeke, S. 1999. "The Vanishing Sacred Groves in Igboland: A Case for Immediate Action." *Nigerian Heritage* 8: 130–141.
Orlowska, Izabela, and Peter Klepeis. 2018. "Ethiopian Church Forests: A Socio-Religious Conservation Model under Change." *Journal of Eastern African Studies*, 12(4): 674-695.
Parkin, David. 1991. *Sacred Void.* Cambridge: Cambridge University Press.
Ranger, Terrence. 1988. "African Traditional Religion." In: *The Study of Religion*. Edited by Peter Clarke and Stewart Sutherland, 106–114. London: Routledge.

—. 1999. *Voices from the Rocks*. Bloomington: Indiana University Press.
Rappaport, Roy. 1999. *Ritual and Religion in the Making of Humanity*. Cambridge: Cambridge University Press.
Robertson, A. and W. Luke. 1993. *The Vegetation and Conservation Status of Kaya Coastal Forests in Kenya*. Nairobi: National Museums of Kenya.
Schaaf, T. 1998. "Sacred Groves in Ghana: Experiences from an Integrated Study Project." In: *Conserving the Sacred for Biodiversity Management*. Edited by P. Ramakrishnan, K. Saxena, and U. Chandrashekara. Enfield, N.H.: Science Publishers. 145–150.
Sheridan, Michael. 2002. "'An Irrigation Intake is Like a Uterus': Culture and Agriculture in Precolonial North Pare, Tanzania." *American Anthropologist* 104(1): 79–92.
—. 2004. "The Environmental Consequences of Independence and Socialism in North Pare, Tanzania, 1961–88." *Journal of African History* 45(1): 81–102.
—. 2008. "The Dynamics of African Sacred Groves: Ecological, Social, and Symbolic Processes." In: *African Sacred Groves: Ecological Dynamics and Social Change*. Edited by Michael Sheridan and Celia Nyamweru. Oxford: James Currey. 9–41.
—. 2016. "Cultural Capital and Structural Power in African Landscapes: The Social Dynamics of Sacred Groves." In: *Cultural Landscape Heritage in Sub-Saharan Africa*. Edited by John Beardsley. Washington, DC: Dumbarton Oaks. 237-261.
—. In press. "When Rain is a Person: Rainmaking, Relational Persons, and Post-human Ontologies in Sub-Saharan Africa." In: *The Social and Cultural Dimension of Climate Change in Sub-Saharan Africa*, Jörn Ahrens and Ernst Halbmayer. Eds. London: Routledge.
—. 2023. *Roots of Power: The Political Ecology of Boundary Plants*. New York and London: Routledge.
Sibanda, Backson. 1997. "Governance and the Environment: The Role of African Religion in Sustainable Utilization of Natural Resources in Zimbabwe." *Forests, Trees and People Newsletter* 34: 27–31.
Spear, Thomas. 1978. *The Kaya Complex*. Nairobi: Kenya Literature Bureau.
—. 2003. "Neo-traditionalism and the Limits of Invention in British Colonial Africa." *Journal of African History* 44(1): 3–27.
Spivak, Gayatri. 2006. *In Other Worlds*. London: Routledge.
Tanganyika African National Union (TANU). 1972. *Siasa ni Kilimo* [*Politics is Agriculture*]. Dar es Salaam: Government Printer.

Tankou, Christopher, et al. 2014. "Variation in Plant Biodiversity across Sacred Groves and Fallows in Western Highlands of Cameroon." *African Journal of Ecology* 52(1): 10-19.

Tchamie, Thiou. 2000. "Évolution de la Flore et de la Végétation des Bois Sacrés des Massifs Kabyè et des Régions Environnantes (Togo)." *Lejeunia* 164: 1–36.

United Nations Development Programme (UNDP). 2004. *Jachie and Kagyasi Sacred Grove Conservation Project*. UNDP Small Grants Program. New York. Accessed December 8, 2022. https://sgp.undp.org/spacial-itemid-projects-landing-page/spacial-itemid-project-search-results/spacial-itemid-project-detailpage.html?view=projectdetail&id=5467.

United Nations Educational, Cultural, and Social Organization (UNESCO). 2005. "Conserving Cultural and Biological Diversity: The Role of Sacred Natural Sites and Cultural Landscapes." Geneva: UNESCO.

United Nations Trusteeship Council. 1955. *Official Records of the 15th Session of the Trusteeship Council 1955.* Supplement 1, Resolutions 25 January–28 March. New York: United Nations.

United Republic of Tanzania. 1997. *Tanzania Forest Policy*. Dar es Salaam.

von der Decken, Carl Claus, and Otto Kersten. 1978 [1871]. *Reisen in Ost-Afrika in den Jahren 1859 bis 1865*. Graz, Austria: Akademische Druck und Verlagsanstalt.

von Hellermann, Pauline. 2016. "Tree Symbolism and Conservation in the South Pare Mountains, Tanzania." *Conservation and Society* 14(4): 368-79.

Warren, D. Michael, and Jennifer Pinkston. 1998. "Indigenous African Resource Management of a Tropical Rainforest Ecosystem: A Case Study of the Yoruba of Ara, Nigeria." In: *Linking Social and Ecological Systems*. Edited by Fikret Berkes and Carl Folke, 158–189. Cambridge: Cambridge University Press.

Willis, Justin. 1996. "The Northern Kayas of the Mijikenda: A Gazetteer, and an Historical Reassessment." *Azania* 31: 75–98.

Willis, K., et al. 2004. "How Virgin is Virgin Rainforest?" *Science* 304: 402–3.

Ylhäisi, Jussi. 2000. "The Significance of the Traditional Forests and Rituals in Tanzania: A Case Study of Zigua, Gweno, and Nyamwezi Ethnic Groups." *Silva Carelica* 34: 194–219.

—. 2004. "Indigenous Forests Fragmentation and the Significance of Ethnic Forests for Conservation in the North Pare, the Eastern Arc Mountains, Tanzania." *Fennia* 182(2): 109–132.

CHAPTER VII

CULTURAL AND ECOLOGICAL SERVICES FROM URBAN SACRED GREEN SPACES IN BHUBANESWAR, INDIA

ANTARYAMI PRADHAN, ALISON A. ORMSBY AND MADHUSMITA SAMAL

Urban green spaces are public open spaces in urban areas, government-managed or private, that provide multiple functions and values and are being increasingly recognized by ecologists globally (Gómez-Baggethun and Barton 2013; Andersson et al. 2014; Livesley et al. 2016). These vegetated green spaces in cities include parks, remnant patches of native vegetation, woodlands, street trees, city square/plaza plantings, burial grounds, private gardens, engineered green infrastructure, green roofs, and sports complexes (Aronson et al. 2017; Loram et al. 2007; Haq 2011). With the rising urban population globally, urban green spaces are key places where people encounter nature (Dallimer et al. 2012; Soga and Gaston 2016; Lepczyk et al. 2017).

Urban green spaces provide a range of benefits, which are often referred to as *ecosystem services* (MEA 2005; Ko and Son 2018; Gomez-Baggethun and Barton 2013). These benefits can be enhanced and integrated into urban planning and design (Hansen et al. 2015). The ecosystem services provided by urban green spaces range from conserving biodiversity to supporting physical, psychological, and social well-being of urban people (Dallimer et al. 2012; Shanahan et al. 2015; Ko and Son 2018). The Millenium Ecosystem Assessment (MEA) in 2005 identified four categories of ecosystem services: regulating, provisioning, supporting, and cultural (MEA 2005). Our study focuses on cultural ecosystem services, which are the non-material benefits people get from ecosystems through spiritual enrichment, recreation, and aesthetic experience in nature (Table 1). The provisioning of such cultural services contributes significantly to the quality of life in cities (Gómez-Baggethun and Barton 2013).

Services	**Description**
Aesthetic	Enjoyment of various aesthetics, such as architecture and natural beauty.
Health	Providing space for mental and physical activities such as walking, exercise, yoga, and playing physical games.
Recreational	Facilitating outdoor services for recreation and leisure activities such as relaxing and enjoying the peaceful atmosphere.
Religious	Providing spiritual stability and activities such as worship.
Social	Providing space for meeting friends and new people.

Table 1. Aspects of Cultural Ecosystem Services

As cities expand, urban park managers and ecologists can support preserving and restoring the remnant green spaces in mega-cities of industrializing countries, including India (Lin and Fuller 2013; Nagendra et al. 2013; Soga et al. 2014). The conversion of green spaces into the built environment is one of the major causes of urban green space reduction (Turrini and Knop 2015). Therefore, through protecting or restoring ecosystems, some of this green space can be retained or reclaimed. It is important to monitor and understand the changes in spatial complexity of urban ecosystems (Nor et al. 2017).

Urban green spaces provide cultural services, such as spiritual, religious and aesthetic benefits (Bolund and Hunhammar 1999; Daniel et al. 2012; Milcu et al. 2013; Bertram and Rehdanz 2015). Green spaces provide opportunities for recreation, thus increasing the quality of life of residents (Fuller and Gaston 2009), as well as areas for people to visit, walk, enjoy, and conserve, as well as the benefits of attractive and livable environments for urban residents (Hartig et al. 2014).

Many studies have been conducted on the sacred groves in rural areas of India, including the cultural and ecosystem services that they provide (Bhagwat 2009; Gokhale and Pala 2011; Pradhan and Ormsby 2020; Rath and Ormsby 2020). In contrast to rural sacred groves or sacred forests, which represents native vegetation protected through formal belief systems (Malhotra et al. 2001), urban sacred green spaces (SGS) are formal or informal gardens, lands and trees associated with any place of worship or spiritual recognition (De Lacy and Shackleton 2017). This includes Christian churches, Muslim mosques, and Hindu or Buddhist temples and their associated green spaces (De Lacy and Shackleton 2017; Cooper 1995). Green spaces associated with urban sacred sites may be protected, not because they are directly worshipped, but because they play an integral role in spirituality and are usually respected by all, including residents and visitors who are not religious (Deil et al. 2005).

Temples in Indian cities are examples of SGS in urban areas (Nagendra 2016; Nagendra et al. 2013) and often harbor a wide variety of plant species, including endemics (Gopal et al. 2018). Within sacred sites, specific trees may be worshiped according to formal belief systems. For example, different sacred tree species are worshipped in India and South Africa due to their religious significance (De Lacy and Shackleton 2017). Globally, there are several tree species that represent sacred trees, including the bodhi or peepal tree (*Ficus religiosa*) and the ceiba or kapok tree (*Ceiba pentandra*) (Ormsby 2021a). In many locations globally, these single trees are protected, as in Indian katte shrines. Despite the importance of cultural services provided by urban SGS, they have not been thoroughly examined in urbanized cities of India. Thus, there is a need for investigation of cultural services provided by the urban SGS, which are often managed for different objectives to other formal urban green spaces (Cooper 2012). Our study addressed this need through interviewing visitors to nine Hindu temples with green space in Bhubaneswar, India.

Literature Review

This section provides a review of previous studies focusing on the perception of cultural services within urban green spaces in different parts of the world, specifically Korea, India, and South Africa. However, only a few studies have been conducted in the context of urban SGS. Ko and Son (2018) investigated the cultural ecosystem services in urban green spaces of Gwacheon, Republic of Korea. In this study, the authors interviewed 240 residents of Gwacheon and found that half of people surveyed used the green spaces for walking and hiking. Others used the spaces for activities such as picnics (13%), sporting activities (9.6%), and enjoying nature (5%).

In Bangalore, India, Derkzen et al. (2017) studied the shifts in ecosystem services available in underserved urban areas. This study suggested how urban development affects the well-being of the urban poor, through changes in ecosystem service availability and people's responses to these shifts. They performed 214 household interviews and group mapping sessions in seven urban lake communities in Bangalore, India and summarized three trends in changes near Bangalore's lakes: privatization of land; pollution and habitat degradation; and restoration followed by gentrification (Derkzen et al. 2017).

De Lacy and Shackleton (2017) studied the aesthetic and spiritual ecosystem services provided by urban sacred sites in Grahamstown, a mid-sized city in South Africa. The authors conducted surveys at 30 places of worship and found that visitors valued church gardens because they are reminiscent of beauty created by God, add character to the site, signify tranquility and peace, enhance one's religious experience, promote remembrance of departed ones, and provide reflection space. This study identified a wide diversity of ecosystem services provided by gardens, with social values being more recognized than ecological, and economic services the least (De Lacy and Shackleton 2017). In this study, the authors found that aesthetic appreciation of a garden was significantly related to woody plant species richness, number, and basal area in the garden.

Gopal et al. (2018) studied the biodiversity of sacred sites in an urbanized Indian megacity, Bengaluru (Bangalore). They assessed biodiversity patterns of two types of Hindu places of worship -- temples and kattes -- and underlying environmental parameters. The authors found that urban sacred sites allow people to benefit from biodiversity and documented that temples provide favorable conditions as sites for urban biodiversity, and thus play an important role in the urban green infrastructure of growing megacities.

Taylor et al. (2020) investigated what visitors want from urban parks. Through interviews, they explored why people visit parks in cities in Australia and New Zealand, and what influenced visitation by understanding what park visitors notice, and how being in an urban park makes them feel. The study concluded that visitors valued places where they could interact with nature in a range of ways, for example, parks made them feel relaxed, happy, and/or peaceful.

The rapid urbanization of Indian cities has increased pressure on the conservation of remnant green spaces in the urban landscapes. Conservation of these urban green spaces has become increasingly important for liveable cities due to the range of ecosystem services that they provide for urban inhabitants (Bolund and Hunhammar 1999; Haase et al. 2014).

Research Goal

The goal of our research was to quantify the role of urban SGS in providing ecosystem services, particularly cultural services, in Bhubaneswar, Odisha. The city has undergone rapid urbanization in the last decade (Swain et al. 2017), negatively affecting biodiversity and associated ecosystem services. Our study documented visitors' perceptions of benefits from urban SGS in Bhubaneswar. We had the following aims:

1. To document the flora and fauna observed in selected urban sacred green spaces of Bhubaneswar city.
2. To assess the perceived benefits to visitors from urban SGS.
3. To determine the purpose for people's visit to the SGS.
4. To assess what can be done to improve the urban SGS.

Study Area

The study area was comprised of nine selected urban SGS within Bhubaneswar municipal area, specifically Hindu temples associated with vegetated areas (Figure 7.1). Bhubaneswar, the largest city in the state of Odisha, contains many sacred cultural landscapes. Bhubaneswar has a tropical savanna climate, with an annual mean temperature of 27.4°C and rainfall of 1,542 mm (Khatoon et al. 2016).

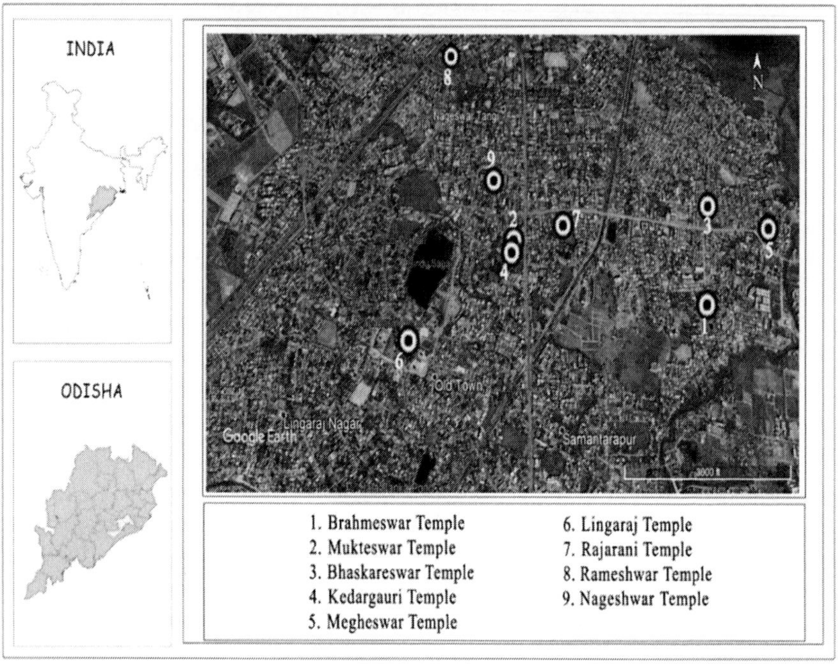

Figure 7.1: Location of Study Sites in Bhubaneswar
Map credit: A. Mansingh

Known as the 'temple city,' Bhubaneswar contains over 500 temples. One example is the Rameshwar Temple, shown in Figure 7.2.

Figure 7.2: Rameshwar Temple, Bhubaneswar
Photo credit: Snigdha Samanwita Sahu

However, not all of the temples are associated with green spaces, and hence only nine temples were selected for this study, specifically those that harbor green spaces around the temple. The sizes of the study site temple grounds are presented in Table 2.

Site Number	Temple Name	Area (hectares)
1	Brahmeswar	1.2
2	Mukteswar	0.5
3	Bhaskareswar	1.9
4	Kedargauri	0.5
5	Megheswar	0.8
6	Lingaraj	0.7
7	Rajarani	1.8
8	Rameshwar	2.4
9	Nageshwar	0.3

Table 2: Sizes of the Study Site Temple Green Spaces

Species Inventory

A tree species inventory was conducted through field visits at nine selected SGS from January through March 2021. Trees were identified using a regional standard flora book (Saxena and Brahmam 1994-96); the names and family affiliations were assigned to the identified species using the online database, The Plant List (2013). All trees and birds were noted while walking for 10-15 minutes along a 50m transect, a method used to survey multiple taxa, such as birds (Watson 2004). While not every species present during the transect walk was recorded due to difficulties in identifying the multiple taxa, the intention was to record as much of the wildlife of the temple grounds as possible, as these can potentially contribute to visitors' attraction towards the urban green space.

Survey Content, Design, and Implementation

From January to March 2021, we invited visitors to temple areas in Bhubaneswar who were at least 18 years old to participate in the interviews, following a questionnaire that we developed. We only asked limited questions in the interview to keep the interaction short and minimize the intrusion on people's temple visit. After verbally agreeing to participate, each participant was asked the following questions:

(1) Why did you visit the urban sacred green space(s)?
(2) What benefits or services do you receive from the sacred green space?
(3) What factors attracted you to visit this place?
(4) What would you suggest to improve the green space?

These questions allowed us to explore the experiences of visitors to the urban SGS and to quantify peoples' perceptions of the benefits they received from the area. Our results also help to understand the factors that attract visitors to these SGS.

Results: Biodiversity of Urban Sacred Green Spaces

The study documented the presence of 39 bird species, which were taxonomically affiliated to 22 families (Table 3).

Common name	Scientific name	Order	Family
Alexandrine parakeet	*Psittaculaeupatria*	Psittaciformes	Psittaculidae
Asian cuckoo	*Eudynamysscolopaceus*	Cuculiformes	Cuculidae
Asian Openbill	*Anastomusoscitans*	Ciconiiformes	Ciconiidae
Asian pied starling	*Gracupica contra*	Passeriformes	Sturnidae
Asian-paradise flycatcher	*Terpsiphone paradisi*	Passeriformes	Monarchidae
Black Drongo	*Dicrurusmacrocercus*	Passeriformes	Dicruridae
Black hooded oriole	*Oriolusxanthornus*	Passseriformes	Oriolidae
Black Kite	*Milvus migrans*	Accipitriformes	Accipitridae
Black-rumped flameback	*Dinopiumbenghalense*	Piciformes	Picidae
Brown headed barbet	*Psilopogonzeylanicus*	Piciformes	Megalaimidae
Cattle egret	*Bubulcus ibis*	Pelecaniformes	Ardeidae
Common myna	*Acridotheres tristis*	Passeriformes	Sturnidae
Common Pigeon	*Columba livia*	Columbiformes	Columbidae
Duck spp.	*Anseranser*	Anseriformes	Anatidae
Greater coucal	*Centropus sinensis*	Cuculiformes	Cuculidae
Grey-bellied cuckoo	*Cacomantispasserinus*	Cuculiformes	Cuculidae
House crow	*Corvus splendens*	Passeriformes	Corvidae
Indian cuckoo	*Cuculusmicropterus*	Cuculiformes	Cuculidae
Indian golden oriole	*Orioluskundoo*	Passeriformes	Oriolidae
Indian pond heron	*Ardeolagrayii*	Pelecaniformes	Ardeidae
Intermediate egret	*Ardea intermedia*	Pelecaniformes	Ardeidae
Jungle babbler	*Turdoides striata*	Passeriformes	Leiothrichidae
Jungle myna	*Acridotheresfuscus*	Passeriformes	Sturnidae
Little cormorant	*Microcarboniger*	Suliformes	Phalacrocoracidae
Oriental magpie robin	*Copsychussaularis*	Passeriformes	Muscicapidae
Pied kingfisher	*Cerylerudis*	Coraciiformes	Alcedinidae
Plain prinia	*Priniainornata*	Passeriformes	Cisticolidae
Purple sunbird	*Cinnyris asiaticus*	Passeriformes	Nectariniidae
Red vented bulbul	*Pycnonotuscafer*	Passeriformes	Pycnonotidae
Red whiskered bulbul	*Pycnonotusjocosus*	Passeriformes	Pycnonotidae
Rose-ringed parakeet	*Psittaculakrameri*	Psittaciformes	Psittaculidae
Rufous treepie	*Dendrocittavagabunda*	Passeriformes	Corvidae
Shikra	*Accipiter badius*	Accipitriformes	Accipitridae
Spotted dove	*Spilopelia chinensis*	Columbiformes	Columbidae
Spotted owlet	*Athene brama*	Strigiformes	Strigidae
Tailor bird	*Orthotomussutorius*	Passeriformes	Cisticolidae
White breasted water heron	*Amaurornisphoenicurus*	Gruiformes	Rallidae
White throated kingfisher	*Halcyon smyrnensis*	Coraciiformes	Alcedinidae

Table 3: Documented Bird Species

The study also identified 39 tree species in the study sites (Table 4).

Tree Scientific Name	**Family**
Artocarpus heterophyllus Lam.	Moraceae
Livistona chinensis (Jacq.) R.Br. ex Mart.	Arecaceae
Mangifera indica L.	Anacardiaceae
Neolamarckia cadamba (Roxb.) Bosser	Rubiaceae
Aegle marmelos (L.) Corrêa	Rutaceae
Psidium guajava L.	Myrtaceae
Ficus religiosa L.	Moraceae
Quercus ilex L.	Fagaceae
Couroupita guianensis Aubl.	Lecythidaceae
Polyalthia longifolia (Sonn.) Thwaites	Annonaceae
Magnolia champaca (L.) Baill. Ex Pierre	Magnoliaceae
Tabebuia aurea (Silva Manso) Bentha. and Hook. f. ex S. Moore	Bignoniaceae
Mimusops elengi L.	Sapotaceae
Tamarindus indica L.	Leguminosae
Saraca asoca (Roxb.) Willd.	Leguminosae
Cocos nucifera L.	Arecaceae
Plumeria alba L.	Apocynaceae
Phyllanthus emblica L.	Phyllanthaceae
Ficus benghalensis L.	Moraceae
Plumeria rubra L.	Apocynaceae
Senna siamea (Lam.) H.S.Irwin & Barneby	Leguminosae
Delonix regia (Hook.) Raf.	Leguminosae
Terminalia catappa L.	Combretaceae
Carica papaya L.	Caricaceae
Annona squamosa L.	Annonaceae
Azadirachta indica A.Juss.	Meliaceae
Prunus serotina Ehrh.	Rosaceae
Ziziphus jujuba Mill.	Rhamnaceae
Alstonia scholaris (L.) R. Br.	Apocyanaceae
Prosopis cineraria (L.) Druce	Leguminosae

Table 4: Tree Species Documented at Study Sites

Visitor Survey Results

Across the nine sites, a total of 325 respondents were interviewed, with varying proportions of males and females depending on age groups (Figure 7.3). The average age was 38, and the respondent ages ranged from 18 to 84 years old.

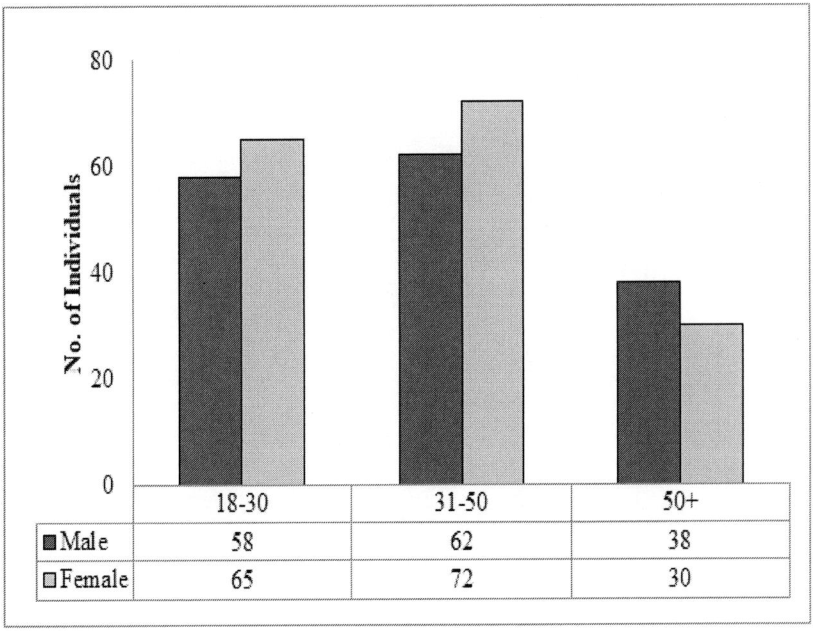

Figure 7.3: Age Groups and Gender Distribution of Respondents

Approximately 35% of the respondents visited the SGS on a daily basis; 4.5% were visiting for the first time. The remainder visited 3-5 times per week (14.5%), 1-2 visits per week (6.5%), 1-2 visits per month (10.5%) and 1-2 visits per year (29%). The respondents visited the SGS from a range of distances. About 48% of the respondents visited the site from a distance of less than one kilometer, followed by 39% from a distance of 1-5km. The remainder, 13% of the respondents, visited from more than 5km.

When asked an open-ended question about what benefits they received from the SGS, visitors identified different cultural ecosystem services. The perceived benefits from the SGS are represented in Figure 7.4 by age group.

The results of the survey show that 45% of the respondents visiting these sacred green spaces received recreational benefits through relaxation and enjoyment of the peaceful and calm environment of the temples. About 27% of the respondents mentioned the contribution of green spaces in providing ecological services like shade, clean air, flowers, and fruits. Despite the religious importance of the site, only about 12% of respondents mentioned the religious or spiritual benefits of these sacred spaces. Other benefits mentioned by the visitors included aesthetic services (7%) and health benefits (8%).

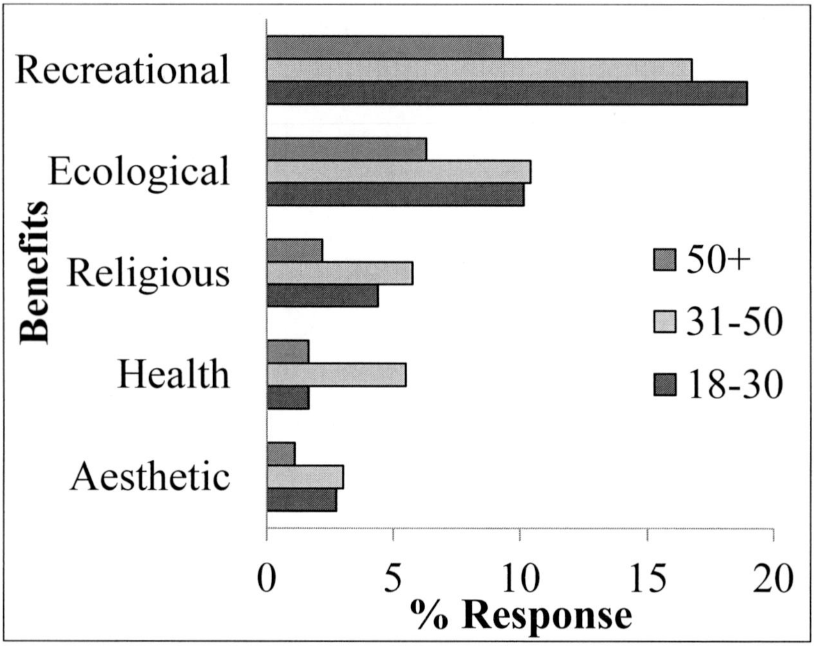

Figure 7.4: Perceived Benefit Indicated by Respondents of Different Age Groups.

Different age groups had different perceptions of the benefits provided from the urban SGS (Figure 7.4). Recreational services were most recognized by the younger age group of 18-30 age (19% of the total response) followed by the 31-50 age group (17%) and least by the older age group with age over 50 (9%). Ecological services were mostly perceived by the middle and older age groups as compared to the younger age groups.

Purpose for Visit to Sacred Green Space

When we asked the visitors about their purpose of visit to these sites, they mentioned different reasons for their visits. The maximum, 33% of the respondents, mentioned visiting for religious purposes, followed by 31.5% of respondents who mentioned physical activities such as yoga, walking, mediation and exercises in the open green spaces. 22% of respondents stated that they visit these sites for relaxation and to enjoy the peaceful atmosphere. About 11% of the respondents visit these sites to enjoy the aesthetic beauty of the ancient temples.

When we compared the cause of visit with the frequency of visit, we found that 67% of the daily visitors and 49% of the regular visitors who come 3-5 times per week were visiting these places for health benefits. Despite the religious importance of these sites, most of the people who visited for religious purpose were occasional (62% of occasional visitors) with only 1 or 2 visits per year. In addition, 81% visitors who visited once or twice a week were coming came for religious purposes.

Suggestions to Improve Quality of Sacred Green Space

The respondents were asked about their suggestions to improve the SGS. Approximately 60% of the respondents suggested that planting be done. Other suggestions, mentioned by 34% of the respondents, related to proper site management, such as recruitment of a gardener, proper care of plants, or irrigation. They also requested that ponds within some of these sacred spaces be restored to attract more tourists.

Discussion

Urban sacred natural sites are gaining increased attention, acknowledging their cultural ecosystem services and their support for biodiversity conservation (Jackson and Ormsby 2017; Gopal et al. 2018; Ormsby 2021b). However, sacred sites in cities are often considered separate from natural areas because of their associated structures like temples and churches (Ishii et al. 2010; Cooper 2012). The cultural ecosystem services provided by urban SGS are particularly important because of their proximity to high concentrations of urban people who appreciate their benefits (De Lacy and Shackleton 2017). Our study measured three key dimensions: the perceived benefits that visitors receive from the urban SGS; the purposes of visits to the SGS; and how to improve the SGS.

We found that almost half of the people interviewed visiting the sacred sites were from nearby places. Taylor et al. (2020) also found the similar observation from park visitors of Australia and New Zealand. This is simply because easy access to the green space was a common reason for the use of green space. Approximately one third of the respondents who we interviewed mentioned religious purpose as their primary cause of visit to the site, as these sites are an important part of the cultural landscape of the city. This was followed by visiting the sites for their health benefits such as for yoga, and to walk or exercise in the green spaces associated with these sites. Similarly, Van den Berg et al. (2010) and Akpinar (2016) documented the improvement in quality of life through the health benefits provided by urban green spaces. Besides cultural importance and health benefits, people in our study also were attracted to the recreational and aesthetic benefits provided by the urban SGS. The results of our study suggest that urban SGS of Bhubaneswar have an important role in providing a wide range of cultural services to urban dwellers.

Conclusion and Recommendations

The findings of the study have the potential to inform the management of existing SGS in cities globally and the establishment of future parks, particularly with sacred spaces, understanding that there are many reasons for people to visit these sacred places. However, the location also plays a role as most of the people visiting these parks for their benefits are due to close proximity of the sacred space. The placement and planning of parks in cities could afford serendipitous interactions with nature and potentially offer opportunities to benefit residents' well-being and encourage more practices to improve biodiversity within green spaces. More investigation is needed to consider the importance of these sites in terms of biodiversity conservation and provisioning of other ecological services. The visitors to the sites we studied pointed out that proper management and planting of native or horticultural plants would positively affect the benefits provided by these places.

The present study of nine urban sacred green spaces can help in framing management policies for future planning and maintenance of these and other sites. More research is needed to explore and restore the ecological and cultural services provided by sacred green spaces in other cities around the world.

Acknowledgements

The authors thank Prof. Niranjan Behera for guidance, and a special thanks to Dr. Debasis Deb by Madhusmita for being on the thesis committee.

Funding:
This work did not receive any grant or other financial support.

References

Akpinar, A. 2016. "How is Quality of Urban Green Spaces Associated with Physical Activity and Health?" *Urban Forestry and Urban Greening* 16. Accessed December 9, 2022. https://doi.org/10.1016/j.ufug.2016.01.011.

Andersson, E., S. Barthel, and S. Borgstorm. et al. 2014. "Reconnecting Cities to the Biosphere: Stewardship of Green Infrastructure and Urban Ecosystem Services." *Ambio: A Journal of the Human Environment* 43(4): 445–453. Accessed December 9, 2022. https://doi.org/10.1007/s13280-014-0506-y.

Aronson, M. F., C.A. Lepczyk, and K.L. Evans et al. 2017. "Biodiversity in the City: Key Challenges for Urban Green Space Management." *Frontiers in Ecology and Environment* 15: 189–196. Accessed December 9, 2022. https://doi.org/10.1002/fee.1480.

Bertram, C. and K. Rehdanz. 2015. "Preferences for Cultural Urban Ecosystem Services: Comparing Attitudes, Perception, and Use." *Ecosystem Services* 12: 187–199. Accessed December 9, 2022. https://doi.org/10.1016/j.ecoser.2014.12.011.

Bhagwat, S.A. 2019. "Ecosystem Services and Sacred Natural Sites: Reconciling Material and Non-material Values in Nature Conservation." *Environmental Values* 18: 417-427. Accessed December 9, 2022. https://doi.org/10.3197/096327109X12532653285731.

Bolund, P. and S. Hunhammar. 1999. "Ecosystem Services in Urban Areas." *Ecological Economics* 29: 293–301. Accessed December 10, 2022. https://doi.org/10.1016/S0921-8009(99)00013-0.

Cooper, N.S. 2012. "The History of English Churchyard Landscapes Illustrated by Rivenhall, Essex." In: Pungetti, G., G. Oviedo, and D. Hooke. eds. *Sacred Species and Sites: Advances in Biocultural Conservation.* Cambridge: Cambridge University Press. pp. 97–110.

Cooper, N.S. 1995. "Wildlife Conservation in Churchyards: A Case-study in Ethical Judgements." *Biodiversity and Conservation* 4: 916–928.

Dallimer, M., J.R. Rouquette, and A.M.J. Skinner et al. 2012. "Contrasting Patterns in Species Richness of Birds, Butterflies and Plants along Riparian Corridors in an Urban Landscape." *Diversity and Distributions* 18: 742–753.

Daniel, T. C., A. Muhar, A. Arnberger et al. 2012. "Contributions of Cultural Services to the Ecosystem Services Agenda." *PNAS* 109(23): 8812-8819. Accessed December 9, 2022. https://doi.org/10.1073/pnas.1114773109.

De Lacy, P. and C.M. Shackleton. 2017. "Aesthetic and Spiritual Ecosystem Services Provided Urban Sacred Sites." *Sustainable* 9: 1628. Accessed December 9, 2022. https://doi.org/10.3390/su9091628.

Deil, U., H. Culmsee, and M.A. Berriane. 2005. "Society's Conservation of Nature for Spiritual Reasons. Sacred Groves in Morocco." *Silva Carel* 49: 185–201.

Derkzen, M. L., H. Nagendra, A.J.A. Van Teeffelen, A. Purushotham and P.H. Verburg. 2017. "Shifts in Ecosystem Services in Deprived Urban Areas: Understanding People's Responses and Consequences for Well-being." *Ecology and Society* 22(1): 51. Accessed December 9, 2022. https://doi.org/10.5751/ES-09168-220151.

Fuller, R.A. and K.J. Gaston. 2009. "The Scaling of Green Space Coverage in European Cities." *Biology Letters* 5: 352–355. Accessed December 10, 2022. https://doi.org/10.1098/rsbl.2009.0010.

Gokhale, Y. and N.A. Pala. 2011. "Ecosystem Services in Sacred Natural Sites (SNSs) of Uttarakhand: a Preliminary Survey." *Journal of Biodiversity* 2(2): 107-115. Accessed December 10, 2022. https://doi.org/10.1080/09766901.2011.11884731.

Gómez-Baggethun, E. and D.N. Barton. 2013. "Classifying and Valuing Ecosystem Services for Urban Planning." *Ecological Economics* 86: 235–245. Accessed December 10, 2022. https://doi.org/10.1016/j.ecolecon.2012.08.019.

Gopal, D., von der Lippe M., and I. Kowarik. 2018. "Sacred Sites as Habitats of Culturally Important Plant Species in an Indian Megacity." *Urban Forestry and Urban Greening* 32: 113–122. Accessed December 10, 2022. https://doi.org/10.1016/j.ufug.2018.04.003.

Haase, D., N. Frantzeskaki, and T. Elmqvist. 2014. "Ecosystem Services in Urban Landscapes: Practical Applications and Governance Implications." *Ambio* 43: 407–412. Accessed December 10, 2022. https://doi.org/10.1007/s13280-014-0503-1.

Hansen, R., N. Frantzeskaki, and T. McPhearson et al. 2015. "The Uptake of the Ecosystem Services Concept in Planning Discourses of European and American Cities." *Ecosystem Services* 12: 228–46.

Haq, S.M.A. 2011. "Urban Green Spaces and an Integrative Approach to Sustainable Environment." *Journal of Environmental Protection* 2(5): 601-608.

Hartig T., R. Mitchell, S. De Vries, and H. Frumkin. 2014 (2015). "Nature and Health." *Ann. Rev. Public Health* 35: 207–228.

Ishii, H.T., T. Manabe, and K. Ito. 2010. "Integrating Ecological and Cultural Values toward Conservation and Utilization of Shrine/ Temple Forests as Urban Green Space in Japanese Cities." *Landscape Ecology and Engineering* 6: 307–315.

Jackson, W., and A.A. Ormsby. 2017. "Urban Sacred Natural Sites–a Call for Research." *Urban Ecosystems* 20(3): 675–681. Accessed December 10, 2022. https://doi.org/10.1007/s11252-016-0623-4.

Khatoon, A., A. Mohapatra, and K.B. Satapathy. 2016. "Fungi Associated with Storage Rots of *Colocasia esculenta L. Tubers* in Bhubaneswar City, Odisha." *British Microbiology Research Journal* 12(3): 1-5.

Ko, H., and Y. Son. 2018. "Perception of Cultural Ecosystem Services in Urban Green Spaces: A Case Study in Gwacheon, Republic of Korea." *Ecological Indicators* 91: 299–306. Accessed December 10, 2022. https://doi.org/10.1016/j.ecolind.2018.04.006.

Lepczyk, C.A., F. La Sorte, and M. Aronson. 2017. "Global Patterns and Drivers of Urban Birds." In: E. Murgui and M. Hedblom. Eds. *Ecology and Conservation of Birds in Urban Environments*. Accessed December 10, 2022. https://doi.org/10.1007/978-3-319-43314-1_2. pp. 13–33.

Lin, B.B., and R.A. Fuller. 2013. "Sharing or Sparing? How Should We Grow the World's Cities?" *Journal of Applied Ecology* 50: 1161–1168. Accessed December 10, 2022.
https://doi.org/10.1111/1365-2664.12118.

Livesley, S. J., A. Ossola, and C.G. Threlfall. 2016. "Soil Carbon and Carbon/ Nitrogen Ratio Change under Tree Canopy, Tall Grass, and Turf Grass Areas of Urban Green Space." *J. Environ. Qual.* 45: 215–223. Accessed December 10, 2022.
https://doi.org/10.2134/jeq2015.03.0121.

Loram, A. et al. 2007. "Urban Domestic Gardens (XI), Variation in Urban Wildlife Gardening in the United Kingdom." *Biodivers Conserv* 16: 3227–38. Accessed December 10, 2022.
https://doi.org/10.1007/s10531-007-9174-6.

Malhotra, K.C., Y. Gokhale, S. Chatterjee, and S. Srivastava. 2001. *Cultural and Ecological Dimensions of Sacred Groves in India*. New Delhi: Indian National Science Academy.

Milcu, A., J. Hanspach, D. Abson, and J. Fischer. 2013. "Cultural Ecosystem Services: A Literature Review and Prospects for Future Research." *Ecol. Soc.* 18. Accessed December 10, 2022. https://doi.org/10.5751/ES-05790-180344.

Millennium Ecosystem Assessment (MEA). 2005. *Ecosystems and Human Well-Being: Synthesis*. Washington, DC, USA: Island Press, p. 155.

Nagendra H., H.S. Sudhira, and M. Katti et al. 2013. Sub-regional Assessment of India: Effects of Urbanization on Land Use, Biodiversity and Ecosystem Services. In: T. Elmqvist, M. Fragkias, and J. Goodness. Eds., *Urbanization, Biodiversity and Ecosystem Services: Challenges and Opportunities*. Netherlands, Dordrecht: Springer. Accessed December 10, 2022. https://doi.org/10.1007/978-94-007-7088-1_6. pp. 65–74.

Nagendra, H. 2016. *Nature in the City: Bengaluru in the Past, Present and Future*. New Delhi: Oxford University Press.

Nor, A.N.M., R. Corstanje, J.A. Harrisa, and T. Brewer. 2017. "Impact of Rapid Urban Expansion on Green Space Structure". *Ecological Indicators* 81(2017): 274-284. Accessed December 10, 2022. https://doi.org/10.1016/j.ecolind.2017.05.031.

Ormsby, A.A. 2021a. "Sacred Groves of Sierra Leone: Preserving Indigenous Environmental Knowledge." In: T.F. Thornton and S.A. Bhagwat. Eds. *The Routledge Handbook of Indigenous Environmental Knowledge*. NY: Routledge. pp. 214-223.

Ormsby, A.A. 2021b. "Diverse Values and Benefits of Urban Sacred Natural Sites." *Trees, Forests and People*. Accessed December 10, 2022. https://doi.org/10.1016/j.tfp.2021.100136.

Pradhan, A., and A.A. Ormsby. 2020. "Biocultural Conservation in the Sacred Forests of Odisha, India". *Environmental Conservation* 47: 190–196. Accessed December 10, 2022. https://doi.org/10.1017/S0376892920000181.

Rath, S., and A.A. Ormsby. 2020. "Conservation through Traditional Knowledge: A Review of Research on the Sacred Groves of Odisha, India." *Human Ecology*. Accessed December 10, 2022. https://doi.org/10.1007/s10745-020-00173-1.

Saxena, H.O., and M. Brahamam. 1994-96. *The Flora of Orissa*. Volumes I – IV. Orissa: RRL, Bhubaneswar and Forest Development Corporation Publication.

Shanahan, D.F., B.B. Lin, and K.J. Gaston. 2015. "What is the Role of Trees and Remnant Vegetation in Attracting People to Urban Parks?" Report 30: 153. Accessed December 10, 2022. https://doi.org/10.1007/s10980-014-0113-0.

Soga, M., and K. J. Gaston. 2016. "Extinction of Experience: The Loss of Human-nature Interactions." *Front Ecol Environ* 14: 94–101. Accessed December 10, 2022. https://doi.org/10.1002/fee.1225.

Soga, M., Y. Yamaura, and S. Koike. 2014. "Land Sharing vs. Land Sparing: Does the Compact City Reconcile Urban Development and Biodiversity Conservation?" *J. Appl. Ecol.* 51(2014): 1378–1386. Accessed December 10, 2022. https://doi.org/10.1111/1365-2664.12280.

Swain, D., G.J. Roberts, J. Dash, and K. Lekshmi et al. 2017. "Impact of Rapid Urbanization on the City of Bhubaneswar, India." *Proc. Natl, Acad, Sci, India, Sect. A. Phys. Sci.* Accessed December 10, 2022. https://doi.org/10.1007/s40010-017-0453-7.

Taylor, L., E.H. Leckey, P.J. Lead, and D.F. Hochuli. 2020. "What Visitors Want from Urban Parks: Diversity, Utility, Serendipity." *Front. Environ. Sci.* 8: 595620. Accessed December 10, 2022. https://doi.org/10.3389/fenvs.2020.595620,

The Plant List- Version 1.1. 2013. Accessed February 18, 2022. http://www.theplantlist.org/.

Turrini, T. and E. Knop. 2015. "A Landscape Ecology Approach Identifies Important Drivers of Urban Biodiversity." *Global Change Biology* 21(4): 1652–1667. Accessed December 10, 2022. https://doi.org/10.1111/gcb.12825.

Van den Berg, A.E., J. Maas, R.A. Verheij, and P.P. Groenewegen. 2010. "Green Space as a Buffer between Stressful Life Events and Health." *Soc. Sci. Med* 70: 1203–1210. Accessed December 10, 2022. https://doi.org/10.1016/j.socscimed.2010.01.002.

Watson, D.M. 2014. "Comparative Evaluation of New Approaches to Survey Birds." *Wildlife Res* 1–11. Accessed December 10, 2022. https://doi.org/10.1071/wr03022.

Chapter VIII

Safeguarding Sacred Groves in Meghalaya: Issues and Challenges

Parismita Borah

The United Nations Educational, Scientific and Cultural Organization (1996) defines sacred groves as the areas which have 'natural' vegetation within its premises and are essentially preserved through the medium of taboos and sanctions which have spiritual and ecological values. These ecological values are found while associating the traditional sacred groves with wildlife and physical lands such as streams. Sacred groves were seen widely prevalent in and across ancient cultures such as that of the Romans, Greeks, Baltic, Germanic etc. In present times, Asia and Africa have reported abundance of such groves with about 14, 000 documented in India alone. The state of Meghalaya in Northeast India with an area of 22, 429 sq. kms has around one hundred and twenty-five sacred groves with sizes ranging from about 0.01 to 900 hectares. The state represents a part of the Indo-Burma hotspot. These sacred sites form an integral part of the traditional knowledge systems (TKS) prevalent amongst the different traditional communities of the state of Meghalaya.

India is very rich in biodiversity and has an estimated thirty-six hotspots. However, the total geographical area covered by forests is decreasing as the data provided by the Forest Survey of India (FSI) in 2005, shows that only 23.8% of the land is covered by forests which is lower when compared to the data provided in the Forest Policy of 1988. The sacred groves in Meghalaya should be understood as part of the traditional knowledge systems through which the indigenous communities of the state have been conserving the environment and its resources. There are various rites and rituals performed in these groves. The people follow certain social norms such as the taboo in relation to cutting down of trees, plucking or touching leaves or breaking twigs which have essentially helped in the sustenance of the ecological resources. The deities are worshipped regularly and strict prohibitions are in place to ensure that the sacred site is not disturbed by human populace. It is important to mention that these patches of vegetation are providers of livelihood for the communities residing around them and hence act as custodians to safeguard the forests tracts through generations from time immemorial. The Khasis, Garos and Jaintias inhabiting the state follow a matrilineal system and this is reflected in the participation of women in the religio-cultural activities related to the sacred groves. Thus, in Meghalaya, it is common to find women performing various rites and rituals associated with these forests. This is in sharp contradiction to the neighbouring state of Assam where the Karbi and Mising women, for example, are forbidden from participating in rituals related to appease forest deities.

Objectives and Methodology

This study attempts to examine the sacred groves in Meghalaya as repository of flora and fauna in the light of the issues and challenges faced in safeguarding the threatened ecosystem. It will also gain insights into the role played by the government for biodiversity conservation and management of the forests in the state. An important dimension of the discussion is the direct role of women in the sustenance of the rich ecosystem which is rarely found in India or across the globe. This is on account of the matrilineal system and implications of gender role prevalent among the major ethnic groups in Meghalaya.

Literature Review

A review of select existing material on the subject under discussion will help to understand the research gap and importance of undertaking the present study.

The book titled, *Sacred Groves in India: An Overview* (K.C. Malhotra, Yogesh Gokhale, Sudipto Chatterjee and Sanjiv Srivastava, 2007) gives a detailed analysis of the sacred groves in India. Sacred groves have grown in significance over the due course of time for their role in safeguarding and protecting the environment, and also for its valuable resources. In Meghalaya the sacred groves have been an age-old tradition.

Sacred Groves and Local Gods: Religion and Environmentalism in South India (Eliza F. Kent, 2013) is another useful work on sacred groves with special reference to the South Indian region. In this work, the author discusses how religion and environment in South India are interlinked and how the local gods form an integral part of the daily activities of the people. This association has given religious and cultural meaning to the forests which has in turn contributed to protection of the groves from degradation.

Conserving the Sacred for Biodiversity Management (P.S. Ramakrishnan, K.G. Saxena, and U.M. Chandrashekhara, 1998) is a collection of articles on sacred groves which discusses a variety of perspectives such as ecological, social, anthropological and cultural aspects.

B.K. Tiwari, S.K. Barik and R.S. Tripathi has studied the sacred groves in Meghalaya in an article titled, "Biodiversity Value, Status, and Strategies for Conservation of Sacred Groves of Meghalaya, India." (*Ecosystem Health, 1998).* The work is an overview of the sacred groves found in the state of Meghalaya. It discusses the age-old practices followed by the indigenous communities and explains how it has been an integral part of their culture.

Cultural Landscape

The major tribes inhabiting the state of Meghalaya are the Khasis, the Garos and the Jaintias and they constitute around 85.53% of the total population of the state. These communities have been closely associated with safeguarding and protecting the sacred groves of the state. The settlement pattern of the tribes in the state are that the Khasis mainly inhabit the central region, the Garos inhabit the western region and the Jaintias inhabit the eastern part of the state. The three tribes follow a matrilineal system which trace their descent through the mother, and where the youngest daughter inherits the property from her mother. The Khasis are subtribes of a larger group known as *Hynniew trep,* translated as, 'the seven huts' in the Khasi language. They are regarded as the earliest settlers in the Indian sub-continent, speaking the Mon-Khmer language and belonging originally to the Austro-Asiatic linguistic group. They had their indigenous religious belief system known as *Ka Niam Khasi* or *Ka Niam Tre* and some of the gods they worship include *U Blei Nong-thaw, Ulei Longspah, Chnong*. However, with the advent of the Christian missionaries, majority of the Khasis today have embraced Christianity. The main occupation of the Khasis is *jhum* or shifting cultivation. Rice is the staple food and they also consume fish and meat. Rice beer is a very popular drink among the Khasis. They celebrate various festivals, most important of which is known as the *Nongkrem*, which is a thanksgiving occasion for the good harvest they reaped and a time when they seek the blessings and protection of the Almighty for the welfare of the community.

The Garos are the second largest community inhabiting the state of Meghalaya comprising around one-third of the total population. They inhabit the Garo Hills district although a small section of them is settled in some districts of Assam and West Bengal. The Garos are said to be a Tibeto-Burman group of people and they believe that Tibet is their place of origin. The Garo people call themselves Achik Mande which translates as 'hill people' or 'people of the hills.' *Songsarek* is the traditional religion of the Garos although like the Khasis majority of the tribe have converted to Christianity. Like the Khasis, the Garo women hold a relatively high status in the society, and the youngest daughter inherits the property owned by the mother. The main occupation of the Garos is *jhum* or shifting cultivation and the main crops include rice, millets, chillies, cotton, ginger etc. The main festival of the Garos is *Wangala* which is celebrated after harvesting of crops as thanksgiving to their deity called *Salijong.*

The Jaintias who inhabit the Jaintia Hills district of Meghalaya, and some districts of Assam may be categorised into three main subtribes, viz; Pnars, Wars and the Bhois. Agriculture is the main occupation of the Jaintias and many festivals are celebrated to offer prayers to nature such as rivers, mountains and peaks. The people consider Kopili as the most sacred river and it is generally believed that in the ancient times, human sacrifices were also made to please the river deity. The *Raliang Daloiship* festival and other fertility rites, and performance of the *Longhai dance* exhibits the merry-making of the whole community on the occasion of good harvest.

Traditional Knowledge and Biodiversity Conservation

The Khasi and the Jaintia Hills districts are rich biodiversity hotspots. Under the traditional forest classification pattern and later adopted by the district council, there are three types of forests which are termed as sacred forests/ groves. These forests are known as *Law Lyngdoh, Law Kyntang* and *Law Niam*. There are many deities which are believed to dwell in the sacred groves and the prominent are U Basa, Ryngkew, and Labasa. Among the 125 sacred groves discovered so far in the state, the two most well-known groves are the Mawphlang sacred grove and the Cherrapunji sacred grove.

The Mawphlang sacred grove, also known as *Law Lyngdoh* sacred grove is considered to be the largest sacred grove in the state. The grove got its name from the *Lyngdoh Mawphlang* clan to whom the forest belonged. In fact, the priestly and the ruling clan of Mawphlang dedicated the sacred site to the guardian spirit whom they call *U Ryngkew U Basa.*

The state of Meghalaya enjoys high rainfall and local climatic variations which have facilitated the growth of rich green vegetation on a large scale. The sacred groves in the state are home to around fifty endangered and about two thousand plant species, the growth of which are particularly confined to the pristine forests. A total of an approximate one hundred varieties of rare plants were reported from the Garo Hills district, out of which around fifty species were endemic. The Jaintia Hills district reported around thirty rare species of plants out of which twenty species were endemic. The value of the sacred forests is not confined only to religio-cultural aspects alone, but has been a repository of traditional knowledge systems playing a major role in sustainable environmental conservation of the state. The ethno-medicinal plants have been used to treat diseases and other health issues, and the sacred groves provide the communities dwelling within its vicinity with medicinal plant species. *Orchidaceae*, a flowering plant of the orchid family is the most widely found species in the Lyngiong sacred forests, located about 30 kms from Shillong, the capital of Meghalaya. Other species such as *Poaceae* and *Leguminosae* are also found within the area of the sacred groves. The sacred groves of the state is also home to certain angiosperms and various rare plant species belonging to ancient plant families such as *Magnoliaceae, Digneriaceae, Eupomataceae, Himantandraceae*; and many others are widely evident in the sacred forests of the state (Jeeva et al. 2006, 564). A list of ethno-medicinal plants found in the sacred groves of the state of Meghalaya are given below:

Botanical Name	Family	Useful parts	Uses
Acanthus leucostachyus Wall.	Acanthaceae	Leaves	Fever, toothache
Achyranthes aspera Linn.	Amaranthaceae	Leaves	Cuts and wounds
Ageratum conyzoides Linn.	Asteraceae	Leaves	Snakebite
Anacolosa crassipes Kurz.	Olacaceae	Leaves	Skin disease
Aporosa octandra A.R.Vickerg	Euphorbiaceae	Bark	Colic fever
Aristolochia cathcartii Hook.f.	Aristolochiaceae	Rhizome	Stomach pain
Artemisia nilagirica Pamp.	Asteraceae	Leaves	Asthma
Artemisia vulgaris Linn.	Asteraceae	Leaves	Headache
Asparagus racemosus Willd.	Liliaceae	Tubers	Fever
Begonia sp.	Begoniaceae	Whole plant	Stranguary
Centella asiatica (Linn.) Urban	Apiaceae	Leaves	Brain tonic
Costus speciosus (Koeing)Sm.	Zingiberaceae	Roots	Fever
Curcuma aromatica Salisb.	Zingiberaceae	Tubers	Gastric troubles
Dicliptera roxburghiana Nees.	Acanthaceae	Roots	General tonic
Disporum calcaratum D.Don.	Liliaceae	Tubers	Eye troubles
Ecbolium linneanum Kurz.	Acanthaceae	Roots	Jaundice
Eclipta alba (Linn.) Hassk.	Asteraceae	Leaves	Hair tonic
Elephantopus scaber Linn.	Asteraceae	Whole plant	Abortion
Emilia sonchifolia DC.	Asteraceae	Leaves	Eye troubles
Eupatorium cannabinum Linn.	Asteraceae	Leaves	Fever
Eupatorium odoratum Linn.	Asteraceae	Leaves	Anticoagulant
Garcinia sopsopia Mabb.	Clusiaceae	Leaves	Antidote for poison
Hedyotis scandens Roxb.	Rubiaceae	Leaves	Cough and cold
Impatiens racemosa DC.	Balsaminaceae	Leaf & root	Rheumatism
Jasminium lanceolarium Roxb.	Oleaceae	Leaves	Skin diseases
Lasia spinosa Thw.	Araceae	Rhizome	Antidote for poison
Oenanthe stolonifera Wall.	Apiaceae	Leaves	Stomach pain
Oldenlandia diffusa Roxb.	Rubiaceae	Leaves	Eye troubles
Oxalis corniculata Linn.	Oxalidaceae	Leaves	Snakebite
Plumbago zeylanica Linn.	Plumbaginaceae	Root	General debility
Pogostemon parviflorus Benth.	Lamiaceae	Leaves	Headache
Polygonum chinense Linn.	Polygonaceae	Roots	Urinary disorders
Pothos kunstleri Hookk. f.	Araceae	Leaves	Antidote for poison
Rhaphidophora hookeri Scott	Araceae	Leaves	Antidote for poison
Rubia cordifolia Linn.sensu Kook f.	Rubiaceae	Root &stem	Eye & ear problems
Rubus khasianus Cardot	Rosaceae	Roots	Cuts and wounds
Smilax prolifera Roxb.	Liliaceae	Leaves	Labour pain
Solanum khasianum C.B. Clark emend.	Solanaceae	Fruits	Toothache
Strobilanthes scaber Nees.	Acanthaceae	Leaves	Skin diseases
Terminalia chebula Retz.	Combretaceae	Fruits	Stomach problems
Viola diffusa Ging.	Violaceae	Leaves	Headache

Table 1: Select List of Ethno-medicinal Plants found in Sacred Groves, Meghalaya

Source: Jeeva et al. 2006. "Traditional Knowledge and Biodiversity Conservation in the Sacred Groves of Meghalaya." *Indian J Traditional Knowledge* 5(4): 565. October.

The above list shows how the people belonging to the indigenous communities heavily rely on these forests to cure various diseases and health related issues. These traditional plants are also known to be used for the manufacture of modern medicines invented by scientists across the world, thus enhancing the need to protect the sacred natural sites. This is linked to collection of the resources and livelihood of the indigenous communities who have the access to these designated hotspots.

It is crucial to note that the ecosystem require a balance of flora and fauna which implies that the natural habitat of animals must be protected as done for the plants. In Meghalaya, out of a total of more than one hundred sacred groves documented so far, around sixty of them are located at the source of water streams. The *Lum Shyllong-Nongkrim* sacred grove is the source of around ten perennial streams. Similarly, the *Mawsmai falls* and the *Maswmai cave* are situated within the area of the *Mawlongsiem* sacred grove which receives very high rainfall throughout the year. The favourable climatic conditions nurture the different species in the natural sites. The faunal species such black bears, elephants, leopards, tigers etc. live in thick forests. The sacred groves of the state are home to the rufous-necked hornbill, a globally threatened species of conservation. In the case of the hornbill species including the great Indian hornbill and Wreathed hornbill, their habitat requires tall and matured trees for nesting and which also acts as a protective hedge for their young ones. Such tall trees are found abundantly within the area of the sacred groves. The sacred groves not only nurture different species within it, but also play a significant role in the conservation of soil and water. As majority of the sacred groves are found alongside water streams which make the soil of these groves very rich in nutrients; it provides a fertile ground for the growth and development of the different forest trees and plant species. A list of certain faunal species found in the sacred forests of Meghalaya is given below:

Zoological Name	Common Name
Cuon alpinus Pallas	Wild dog
Elephas maximus Linn.	Asian elephant
Hylobates hoolock (Harlen)	Hoolock gibbon
*Lutra perspicillata*Geoffray	Smooth-coated otter
Macaca arctoides	Bear macaque
Macaca assamensis M'Clelland	Assamese macaque
Manis crassicaudata Gray	Chinese pangolin
Melursus ursinus Shaw	Sloth bear
Panthera tigris (Linn.)	Tiger
Pardofelis nebulosa	Clouded leopard
Semnopithecus pileatus	Capped leaf monkey
Ursus malayanus	Malayan sun bear
Viverra zibetha	Indian pangolin

Table 2: Select List of Faunal Species in Sacred Groves, Meghalaya

Source: Jeeva et al. 2006. "Traditional Knowledge and Biodiversity Conservation in the Sacred Groves of Meghalaya." *Indian J Traditional Knowledge* 5(4): 567. October.

The richness in the biodiversity and ecology demands for the need to preserve and protect these sacred groves for the protection of the resources and to protect the species from extinction in the near future. The indigenous communities of Meghalaya have been safeguarding these groves as their own from the earlier times and this tradition has passed down through generations till the recent times. The traditional priests called *lyngdohs* have the primary responsibility of protecting these groves from external disturbance. In contemporary times, the district council of Meghalaya has vested the power to the *lyngdohs* to take care of these sacred forests. The priests visit the sacred sites at least once a year and perform rituals which has both religious and ecological significance. It may be noted that the biodiversity and ecological assets of any state or region forms an important part for sustaining life of people and other living organisms.

Sacred Groves: Challenges and Issues

Deforestation is one of the major causes of depleting forest covers worldwide. Unsustainable forest management and human activities in the name of development is threatening the world's habitable lands. In this context, preserving the ancient and age-old sacred forests becomes a challenge. As such the number of sacred groves is also decreasing in recent times which may lead to extinction of invaluable resources. In the context of Meghalaya there has been an increase in the tendency to use the resources available in these sacred groves for both individual consumption and commercial purposes. The most important resource derived from the sacred groves of Meghalaya is timber and woods of valuable trees which are used as firewood or for commercial purposes. There has been unabated illegal encroachments in the reserved forests and this intrusion by humans disturbs the habitat of the animals. Many rare species of animal are also seen to be killed to be sold in the international market for deriving a huge sum of money against the same. Although, the indigenous communities have been trying their best to protect these areas from outer illegal intervention, it is alleged that in some cases it is evident that certain individuals belonging to these communities themselves get into the process of allowing illegal entry to these groves in return of certain amount of bribe. A factor of significance is the history of the communities which has seen changes in the religion and cultural fields. There is a view that having adopted new faiths, there is a lesser understanding of the traditional belief systems which has posed serious threats to the sacred groves of the state. Ritual practices had the function of reviving and being in contact with the natural vegetation sites. The *Mawphlang* and *Raliang* sites have been witnessing the practice of rites and rituals which have passed down from the earlier generations. The weakening of the traditional pattern of belief systems and practices has been also evident in the other parts of India in states such as Karnataka and Kerala which are rich areas of sacred groves. This, in turn has posed threat to the protection and conservation of the age-old tradition or notion of sacred groves.

With these increasing threats to the sacred forests, it becomes very important to take up necessary measures and initiatives to protect and preserve these valuable biodiversity rich areas. The indigenous communities have a major role to play in this regard with the active support and cooperation of the state government. Only then the conservation of these groves can be achieved to a satisfactory level.

Safeguarding Sacred Groves: Role of the State

The government of a state has a crucial role to play in the preservation of the ecosystem of the state on the basis of the global guidelines. In the case of Meghalaya, the government has entrusted the responsibility of protecting the sacred groves to the religious priests or *lyngdohs*. The District Councils register the sacred groves under its supervision and the religious priests are given the full autonomy to protect and safeguard the sacred groves. It is worthwhile to mention that unlike the other states in India, the sacred groves of Meghalaya enjoy an adequate legal support. The sacred groves of the state are covered under the United Khasi-Jaintia Hills Autonomous District (Management and Control of Forests) Act, 1958 and the Garo Hills Autonomous District (Management and Control of Forests) Act, 1961.

The *Constitution of India: Para 11, Sixth schedule under Article 244(2) and 275(1),* has outlined the provisions on how to function or run the administration of the Tribal Areas in "the states of Assam, Meghalaya, Tripura and Mizoram". Under this Act, almost all the forest lands of the state have been constituted except for an area of approximately 8% of the total forest land which are directly under the supervision of the State Forest Department. It is mentioned in the Acts, that the sacred groves of Meghalaya are to be managed by the *lyngdoh* or a person(s) who is regarded as the head of a particular village community or locality. As the maintenance of the forests is governed by the laws framed by the District Council of Meghalaya, the traditional priests are accountable for any illegal activities taking place in the protected area. The documentation of sacred groves under the District Council has been provided in the United Khasi-Jaintia Hills Autonomous District (Management and Control of Forests) Act, 1960. The guidelines framed for conservation mentions that no tree within the vicinity of the forests should be felled without the written permission of the Chief Forest Officer of the District Council or any other officer which he has appointed and given the charge in that context. Further, section 9 of the same Act state that no trees belonging to *Law Lyngdoh, Law Kyntang* and *Law Niam* areas should be felled except for use in religious rites, rituals and ceremonies after the proper recognition and sanctions.

In addition to the mentioned Acts, the North Eastern Council (NEC) which was constituted as a statutory advisory body under the NEC Act of 1971 came into effect on 7th November, 1972 at Shillong acts as a significant organization in shaping the biodiversity of the state of Meghalaya. The Act enables the Council to secure the balanced development of the North-eastern states and to look into, from time to time, the successful implementation of various policies, projects and schemes in regard to the matters which are of common importance to the areas. Since, the state of Meghalaya is seen to share its biodiversity areas with the states of Assam and West Bengal, the NEC Act enables the state to initiate projects and schemes for the development of these regions for which the benefits earned would be shared and the expenditure required for these changes would be given by the Council. The NEC Act of 1971 was further amended with slight modifications within it and renamed as the North Eastern Council (NEC) Act of 2002. This amendment was made with the inclusion of Sikkim as a new state within the Northeastern region of the Indian sub-continent. However, major points in the document remains the same with slight addition with special reference to the state of Sikkim.

With the initiation of the North Eastern Region (NER) vision, 2020, which was unveiled in July, 2008, in Delhi, the biodiversity of the state of Meghalaya seems to face a challenge. The NER 2020 vision aims for development of infrastructure in terms of roads, railways, airways and also development in terms of power and telecom connectivity. The 12th Five Year Plan (2012-17), has documented that with more focus upon the infrastructural development, the north eastern states have witnessed a major boost in the gross domestic product. The gross domestic product was highlighted to be around 9.8% which was higher than the gross domestic product as highlighted in the 11th Five Year Plan which was 8%. The North Eastern Council (NEC) Regional Plan has also been initiated to supplement the developmental efforts of the government for the overall development of the north eastern states.

Within this context, though major initiatives have been taken up by the government for the overall development of the state of Meghalaya, the preservation and conservation of the rich biodiversity and ecological assets remain limited. The approximate areas of land under the traditional sacred groves in the state are seen to be decreasing in the recent times. With infrastructural development, urbanization has been witnessed in the state with the coming up of new industries and other institutions in the name of infrastructural development. Here, what should draw the attention of researchers is regarding the land upon which new infrastructural structures have been built. Recent studies reveal that almost half of the land areas of the sacred groves in the state have been meticulously destroyed due to developmental activities. Only a very few of them remain untouched or undisturbed. Urbanization has led to severe threats to the environment of the state such as low soil quality, increase in air pollution and most significantly many of the indigenous communities who have been residing in the areas adjoining the sacred groves are seen being disturbed regarding their homes and the basic necessities which they collect from these forests for their day-to-day survival.

Discussion and Comments

From the above discussion, certain observations have been made. First, the sacred groves form an integral part of the day to day lives of the indigenous communities living in the state of Meghalaya. They depend on these forests for their livelihood and have been protecting them as their own property since the ancient times. Second, the sacred groves are seen to have been reducing in size in the recent times. Most of these groves have been intervened by commercial factors and have lost its traditional values. Third, with the emerging development in the field of infrastructure and rapid urbanization, the indigenous communities have been seen to lose the practice of traditional norms and customs. Only a very few groves in the state are still practising the customs laid down by the forefathers of these communities. Fourth, though the government has announced policies to safeguard these groves, there is much that is left in terms of implementation. Fifth, these groves are community-preserved ecological areas and as such are not immune to encroachments and other illegal activities.

What follows from the above discussion is that there should be an increased campaign on environmental awareness at the grassroot level to educate all sections of the population. The government should go hand-in-hand with the non-governmental organisations to support the cause of environment. There should be regular interaction between the religious heads and the government in matters relating to these sacred groves. Indeed, there is an urgent need to integrate environmental sustainability and social development if the *Sustainable Development Goals: Agenda 2030* is to be achieved.

Figure 8.1: Mawsmai cave located in Mawlongsiem Sacred Grove, Meghalaya

References

Bhagwat, Shonil A. and Claudia Rutte. 2006. "Sacred groves: Potential for Biodiversity Management." *Frontiers in Ecology and the Environment* 4(10): 519-524.

Gadgil, Madhav, and Vaman D. Vartak. 1976. "The Sacred Groves of Western Ghats in India." *Economic Botany* 30(2): 152-160.

Gadgil, Madhav and M.D. Subash Chandran. 1992. "Sacred Groves." In: *India International Centre Quarterly* 19(½): 183-187.

Forest Department, Ministry of Forest, Environment and Climate Change, Government of Goa. Accessed April 11, 2022. https://forest.goa.gov.in/node/49.

India State of Forest Report (ISFR). 2005. Dehradun: Forest Survey of India, Ministry of Environment, Forest and Climate Change, Government of India.

Jeeva, S., B.P. Mishra, N. Venugopal, L. Kharlikhi, and R.C. Laloo. 2006. "Traditional Knowledge and Biodiversity Conservation in the Sacred Groves of Meghalaya." *Indian Journal of Traditional Knowledge* 5(4) October: 563-567.

Khan, M.L., Ashalata Devi Khumbongmayum, and R.S. Tripathi. 2008. "The Sacred Groves and Their Significance in Conserving Biodiversity: An Overview." *International Journal of Ecology and Environmental Sciences* 34(3): 277-291.

"Khasi Tribe, Garo and Jaintia in Meghalaya." Accessed March 27, 2022. https://easternroutes.com/northeast-india/meghalaya/khasi-tribe-garo-jaintia/.

Kent, Eliza F. 2013. *Sacred Groves and Local Gods: Religion and Environmentalism in South India*. New York: Oxford University Press.

Malhotra, K.C. 1998. "Anthropological Dimensions of Sacred Groves in India: An Overview". In: P.S. Ramakrishnan, K.G. Saxena, and U.M. Chandrashekara. eds., *Conserving the Sacred: For Biodiversity Management.* New Delhi: Oxford and IBH. 423-438.

Malhotra, Kailash C., Yogesh Gokhale, Sudipto Chatterjee, and Sanjeev Srivastava. 2001. *Cultural and Ecological Dimensions of Sacred Groves in India.* New Delhi: Indian National Science Academy, and Bhopal: Indira Gandhi Rashtriya Manav Sangrahalaya.

Nongkynrih, Kynpham Sing. "The Sacred Groves of Meghalaya." Accessed March 12, 2022. https://science.thewire.in/environment/the-sacred-groves-of-meghalaya/.

North Eastern Council Act. 1971. Shillong: North Eastern Council, Government of India.

North Eastern Council (Amendment) Act. 2002. Shillong: North Eastern Council, Government of India.

North Eastern Region Vision. 2020. Shillong: North Eastern Council, Government of India.

North Eastern Council Regional Plan (2017-18 to 2019-20), Draft. Accessed April 27, 2022. https://necouncil.gov.in/about-us/nec-regional-plan-0.

Ormsby, Alison A. 2011. "The Impacts of Global and National Policy on the Management and Conservation of Sacred Groves of India." *Human Ecology* 39(6): 783-793.

Ormsby, Alison. 2013. "Analysis of Local Attitudes Toward the Sacred Groves of Meghalaya and Karnataka, India." *Conservation and Society* 11(2): 187-197.

Rehmani, C. 2012. "Rural India, the Emerging Growth Centre for Indian Retailers." *Indian Streams Res J* 2 (11): 1–3.

Ray, Rajasri, M.D.S. Chandran, and T.V. Ramachandra. 2014. "Biodiversity and Ecological Assessments of Indian Sacred Groves." *Journal of Forestry Research* 25(1): 21-28.

"Sacred Groves." Thiruvananthapuram: Kerala State Council for Science, Technology and Environment. Accessed March 17, 2022. http://www.kerenvis.nic.in/Database/SacredGroves_1433.aspx.

"Sacred Groves." Shillong, Meghalaya: Forest and Environment Department. Accessed April 5, 2022. http://www.megforest.gov.in/forest_sacredgroves.html.

Sixth Schedule of the Indian Constitution: Article 244(2) and Article 275(1). Accessed April 27, 2022. https://www.mea.gov.in › Images › pdf1.

State of Forest Report. 2005. Introduction. Accessed April 11, 2022 https://fsi.nic.in/sfr2005/chapter1.pdf.

The State of the World's Forests. 2020. Food and Agricultural Organisation of the United Nations. Accessed December 13, 2022. https://www.fao.org/contact-us/en/.

The United Khasi-Jaintia Hills Autonomous District Council (Management and Control of Forests) Act, 1958. 2019. Compendium of Acts and Rules of Autonomous District Councils. Volume V. Shillong: Forests and Environment Department, Government of Meghalaya. 5-20.

The United Khasi-Jaintia Hills Autonomous District Council (Management and Control of Forests) Act, 1960. Accessed April 27, 2022. https://www.india.gov.in/united-khasi-jaintia-hills-autonomous-district-act-1960.

Tiwari, B.K., S.K. Barik, and R.S. Tripathi. 1998. "Biodiversity value, status, and strategies for Conservation of Sacred Groves of Meghalaya, India." *Ecosystem Health* 4(1): 20-32.

Tribal Heritage in India Foundation. Accessed March 17, 2022. https://indiantribalheritage.org/?p=14246#gsc.tab=0.

Upadhaya, Krishna, Harendra Nath Pandey, P.S. Law, and Radhey Shyam Tripathi. 2003. "Tree Diversity in Sacred Groves of the Jaintia hills in Meghalaya, Northeast India." *Biodiversity and Conservation* 12(3): 583-597.

CHAPTER IX

FROM SACRED TO SECULAR: SACRED GROVES OF NORTHEAST INDIA IN TRANSITION

MEETA DEKA

Northeast India, where landscape, geography and ecological diversity are interwoven with multiple ethnic communities speaking different Tibeto-Burman and Austric languages practising diverse religions and belief systems, is home to complex and distinct cultures. A symbiotic relationship between nature and its people has adversely impacted the region in diverse ways, climate change and land use in particular, resulting in an encounter with the non-humans and the natural habitat, implying a need to unlearn 'development' in the current era. Though the eight states, viz., Arunachal Pradesh Assam, Manipur, Meghalaya, Mizoram, Nagaland, Tripura and Sikkim, account for 23.75 per cent of the country's total forest cover, the region has witnessed loss of 1,020 square kilometres of forest during 2019-2021 (SFR 2021). Forests still play an integral role in the lives of the people where sacred groves are identified as spaces that allow a reciprocal connection between them and nature. The number of sacred groves in the region have however, largely differed from report to report.

Sacred groves are borne out of diverse historical, ecological, economic, socio- cultural and even political factors that shape a particular culture and hence the concept of 'sacred groves' itself varies the world over. Sacred groves of Northeast India, known by diverse indigenous names, have been in a state of transition, acquiring newer dimensions through various phases of history and hence are not static in time, space and meaning. Throughout the centuries, the connotation of the term has been in a constant state of flux. What was once a secular reverence to the 'power' of nature evolved into a sacred space, only to be undermined under British cultural hegemony and economic exploitation of which the sacred groves of Mizoram are a case in point. Post Independent Northeast India saw a spurt of ethnic movements in two phases, particularly after the Official Language Act of 1960 and later after the Assam Movement (1979-1985 CE) when indigenes feared a political as well as cultural hegemony of the dominant community of Assam and looked upon the sacred groves as an integral part of 'shared pasts', a tangible cultural heritage playing a significant role in the traditional cultures and lifestyles, in asserting their respective ethnic identity. In the twenty-first century and after, with growing concern for global warming, sacred groves of Northeast India assumed the status as a symbol of conservation of biodiversity and ecology and a tourist destination!

Secular Reverence to Nature

A few broad-based definitions of sacred groves by historians, conservationists, environmentalists and others in India raise fundamental questions linked to the historical production of sacred landscapes. An oft quoted definition refers to sacred groves as 'segments of landscape containing vegetation, life forms and geographical features, delimited and protected by human societies under the belief that to keep them in a relatively undisturbed state is expression of an important relationship of humans with the divine or with nature' (Hughes and Chandran,1997). Sacred groves are defined as patches of near natural vegetation dedicated to ancestral spirits or deities, preserved on the basis of religious beliefs and as the best practices of biodiversity conservation of ethnic communities (Amirthalingam 2016; Chandrakanth 1991; Upadhyay et al. 2019; Jeeva et al. 2006; Sukumaran et al. 2018; Hazarika 2015), but has of late 'assumed immense anthropological and ecological importance' Malhotra (2001). Madhav Gadgil (2018), who carried out studies of the sacred groves of Manipur and Mizoram, defines sacred groves in a different perspective: 'a legacy of prehistoric traditions of nature conservation, sacred groves are patches of forest that rural communities in the developing world protect and revere as sacrosanct'. Jeeva and others (2006) emphasize its current significance as,

> Natural museums of giant trees, treasure house of threatened species, dispensary of medicinal plants, regulator of water sheds, recreation centre for urbanites, veritable garden for botanists, gene bank of economic species, paradise for nature lovers and laboratory for environmentalists.

While most definitions see sacred groves as the best traditional practice of religious and cultural beliefs, pure superstitious beliefs, nature worship or even as biodiversity conservation, Malhotra and Jeeva's definition indicates a transition in the connotation of the term formed on the basis of spiritual and religious beliefs to acquiring an ecological significance in recent times. It should be noted that prehistoric tradition of sacred groves could not have been intended as a legacy of nature or environmental conservation nor preserved on the basis of religious beliefs, as many would understand or would like to believe as such today. Gadgil (2018) stated that the ecosystem may be protected through religious beliefs but, is grounded in secular benefits. He debunked the idea of a spiritual or religious association by conservationists, pointing out that if so is the case, then sacred groves should be concentrated at religious sites such as mountain peaks or sources of rivers. His study in the Western Ghats proved that sacred groves need not necessarily be confined to particular habitats. This secular aspect based on a traditional knowledge-based ecosystem, is generally disregarded in most of the definitions.

Culturalization of Nature and Concept of the 'Sacred'

The evolution of sacred groves involves a process of 'culturalization of nature,' a term used by Irmgrad Schultz (2006, 381) though on a completely different context. It is used here to mean how cultural dimensions, whose foundations may be largely secular, are extended to the natural world over time. The concept of 'sacred groves' was perhaps based on the idea that 'power is apprehended through its various manifestations of form' as Gerardus van der Leeuw maintains, 'The sacred, … must possess a form: it must be localizable, spatially, temporally, visibly and audibly or still more simply, the sacred must take place' (Dadosky 2004, 17-18).

Sacred groves perhaps began as a secular reverence to the power of nature rather than 'nature worship' itself; Association with ancestral spirits or deities or any particular religion may have been only a later development. Leeuw provides a different perspective on the primordial categories of human existence. He abstracts the notion of Power and maintains that this notion has universal applicability in respect to primitive cultures (Dadosky 2004,18). He asserts that in the initial state of the human mind, the idea of Power is devoid of any understanding of a deity or even the supernatural. To quote James on Leeuw:

> 'Power appears neither as God nor as what moderns might call the supernatural… For [the primitive mind] … "the naturalness of nature" has yet to be established'. Power is that which is 'highly unusual… [with] no inherent moral value'. Power is the 'highly exceptional and extremely impressive 'Other', 'a departure from all that is usual and familiar'. Power

> of nature may be manifested by natural phenomenon like thunder, lightning, earthquake associated with the forests, that leads to astonishment, wonder, awe and even fear in the humans who encounter them. The primitive mind regards 'such potency of Power as not something that is supernatural but is extraordinary---[that]which we call sacred'.
>
> (James 1995, 223, 244; Dadosky 2004, 17)

Again, Ariel Schwartz (2014) argues that to delineate any space as 'sacred' it requires that it must be different and unique, a break or separate from the rest of space. This naturally involves curious questions: 'Who or what renders a space "sacred"? Through what process? For what duration? Under what terms? … what does it mean for a physical space to be "sacred"?'. Schwartz refers to the classic views of Mircea Eliade and Gerardus van der Leeuw, in determining answers to these questions. Hence in understanding the 'sacredness' in sacred groves, it is essential to not consider 'sacred' and 'holy' as synonymous, or a belief in the supernatural but to associate Fredericke Schleiermacher's concept of *Gefuhl,* 'a feeling of absolute dependence' (Dadosky 2004, 14) on this powerful 'Other', with no strings attached to religion. Amaury De Riencourt (1989, 17) describes the early man as one who stands in 'awe' of all natural phenomena such as storm, lightening, earthquake, volcanic eruption etc. but 'increasingly capable of abstract thought … [and] was afraid of nature's mysterious displays of grandiose power'. Leeuw states that primitive world reveres the power manifested through nature or natural objects and not 'nature' per se, for nature was never conceived as a realm set apart or over against humankind, nor again were objects of nature in principle distinguished as separate from them (Dadosky 2004, 17-18). The Mendha and Kurubas of Nilgiri Hills believe that they are part of a community of beings, comprising not only humans but also mountains, rivers, trees, animals and birds. These nonhuman community members benefit them in many ways; hence, humans must reciprocate through restrained use, even veneration of its 'sacredness' and total protection (Gadgil 2018). This is true of the Dimasas in Northeast India who even include rocks to the above category.

Variations in Sacred Groves

This process of culturalization of nature reveal complex relationships between humans and nature; It is created and shaped by cultural values, personal ties and other determinants and therefore subject to political intervention as well. It is important to understand Nature as Berghoefer (2010) puts it, not 'just a thing that is somewhere "out there," to be either consumed or protected. Instead, it is constituted through a three-way relationship between the individual, society, and the physical world. We understand nature itself not as a given, causal, objective entity, but as a sphere whose boundaries are dynamic and socially constructed' and hence subject to historical change. One needs to look into the societal relationships with nature --- 'The way humans interact with nature shapes their understanding of it, whereas the ways in which nature is represented in turn frame and inspire the ways in which people behave toward and inhabit it. Both elements, society and nature, are what they are only through their relationship to each other' (Berghoefer 2010).

It is in this context that we need to understand the different perspectives on sacred groves apparent in Northeast India based on the different modes of living. Marion Glaser (2006, 127) states that while it is the social systems that create structures of meaning in understanding nature, it is again the 'various cultural lenses which generate different virtual realities of "nature." Berghoefer (2010) and other scholars have pointed out that the distinctive 'dichotomy between use and protection has little explanatory relevance if it is not specified which or whose nature is to be used and from what it is to be protected'. It is not just about resource use only, but also about 'other material interactions, sensory interactions, identity-related aspects, and modes of knowledge acquisition shape people's relationships with nature.'

> Hence, they can inform conservation efforts and provide a differentiated view of local knowledge… [for] many eyes make many natures, and that these many natures constitute a valuable heritage linking the cultural and biological diversity of our world.

The above quote clearly explains the diversity of sacred groves in the region, their varied significance or implications that depended not only on the diverse modes of living but factors in mundane and historical aspects. This diversity is discerned not merely from the names of the groves under various management systems of priests and local councils and the distinctive deities dedicated to them, but also in practice. A striking peculiar trait is found in the sacred groves of the Dimasa, in contrast to most of the other groves. While in most groves plucking of plants or felling of trees is a serious taboo, it has not been so in the Dima Hasao District. *Melocanna buccifera culm* and *Musa sp.* leaf are used in rituals for water purification during Misengba ceremony, Judima (rice beer) is offered in cups made from the culms to the deities while use of a pair of both fresh leaves and Areca catechu fruits (Gowai-DI) is mandatory in most religious practices, and other plants or herbs for medicinal purpose. Twig is used for Maimuhtarba, sacrificial ritual of the buffalo while twig with inflorescences is used in Dakinsa, a religious ceremony, and fresh fruits and certain leaves are used as plates and also considered as a cure for bed-wetting in children. Paste of particular seeds are used for death rituals while dried fruits of a plant are used as decoration over the door pediment (Medhi and Borthakur2013, 71-72). So, there is an extensive use of resources for various purposes from the sacred groves of the Dimasa.

Sacred groves of Manipur are commonly known as *Umanglai* and a ritualistic celebration, *Umanglai Haraoba* of the Meitei community takes place every year. Reports on the number of sacred groves vary; Several are partly threatened, while a few are well protected and preserved (Khumbongmayum, 2004) with acres of 'ecologically important tree species with good soil nutrient retaining capabilities e.g., *Alibizia lebbeck, Ficus resemosa*, etc. reported from these groves' (Khumbongmayum, 2005). These groves dedicated to deities like Umanglai, Ebudhou Pakhangba, Konthoujam Lairembi, Chabugbam and Chothe Thayai Pakhangba are locally known as *gamkhap* and *mauhak* or sacred bamboo reserves where extraction of bamboo shoots is totally prohibited. (Khumbonmayum, 2004; Amirthalingam, 2016).

Darlong and Barik (1998) distinguished between sacred groves and community forest reserves of Northeastern India and broadly classified them into two: *first,* based on religious beliefs and taboos and *second,* in the form of village forest reserves. They describe the former as in Meghalaya and Manipur while the latter is common in Mizoram. Such a distinct classification however maybe problematic. Curiously, the same year, Gadgil and a few other scholars pointed out a new dimension to describe sacred groves '*gamkhap*' of the Gangte, a group of Kuki tribe in Churachandpur district, Manipur, as a resource conservation indigenous practice arrived at through experience, a natural trial-and-error process. What is interesting is that these forests which were initially based on religious beliefs were abandoned consequent to forest fire during burning operations, increased value of timber and conversion to Christianity, but were revived essentially as an ecological conservation practice. Two decades later in 2018, he reiterates that in mid-twentieth century after the destruction of sacred groves during colonial rule and after, several villages like Saichang for example, saw its Gangte community take to 'community-based vigilance and protection' of residual forest patches with the same zest of the past, and even terms them as *gamkhap* or sacred groves but which however are no longer regarded as 'abodes of spiritual beings'. These findings led him to believe that 'even if spiritual beliefs give way to secular ones, sacred groves could survive if ecosystem services reached the ecosystem people'. Gadgil (2018) thus categorically supports the idea that sacred groves could sustain even without the association of religion or spirituality.

Historical Background

One obvious factor that led to the identification of 'sacred groves' *per se*, associated with beliefs or deities, was extension of agriculture following population growth over the centuries. Since very early times, agriculture, rice being the main crop, in Northeast India was based on wet-rice or terrace cultivation and *jhum* or slash and burn, also shifting cultivation, particularly in the hill areas. With increase of population pressure on land, Jhum cycles in the region ranged from 20-30 years in the past to a year in recent times while sequential cropping varied from community to community, leading to degradation of soil and hence, poor economic yield, and insecure livelihood causing an imbalance to the traditional ecosystem (Bhuyan 2019). The destruction of nature that followed since the early period necessitated a need to protect specific parts of the forest through means of cultural association, feelings and emotions, in the form of spirits, superstition or deities to reinforce the power of nature and as an appeal to the masses. Thus began a process of 'culturalization of nature'. Curiously, *jhum* and sacred groves are coexisting contradictory concepts in Northeast India; While *jhum* implied destruction of forests for a particular period, sacred groves meant a kind of conservation integrated into land-use practices. Both traditional practices are associated with nature, indigenous cultural beliefs and practices, superstitions, economy and ecology.

Under British colonialism, by the second half of the 19th century, forests were getting depleted with a vigorous process of resource extraction. This necessitated the British Indian Government to invite German botanist Dietrich Brandis as the first Inspector General of Forests for damage control. Brandis' field study in the early decades of the 20th century mentions large parts of Northeast India like South Lushai Hills, Patkai Range, Mikir Hills of Sibsagar District, Khasi Hills Sylhet, Cachar, Nagaland, Manipur, Sikkim Garo Hills, Chittagong Hills, Mishmi Hills, Upper Assam, Sibsagar and Lakhimpur districts etc. (Brandis 1921). Gadgil (2018) mentioned that Brandis lamented the obliteration of scared groves on one hand, and was highly appreciative of the community-based forest management on the other. He thus suggested that a substantial part of forest be left under the management of the people. The British included this suggestion, only half-heartedly and on a limited scale, in 1930.

Indian Forest Policies that aimed at resource extraction depleted existing forests and undermined the existing sacred groves of the region. The colonial government sought to legitimize their hegemony over the forests of Assam through legislation. It is said that it was Brandis who first facilitated the formulation of the first Indian Forest Act of 1865. The Indian Forest Acts of 1865 and 1878, Assam Forest Regulation 1891, Forest Policy Resolution of 1894, Indian Forest Act 1927 which replaced Act of 1878 deleted the reference to communities' rights over forests, Assam Forest (Amendment) Act, 1931 were some enactments made to extend State control over forests. The Assam Forest Regulation of 1891 brought *jhum* cultivation under state control (reiterated by Indian Forest Act of 1927) (Regulation, 1891, Chapter 2, 11-12) but was silent on the existence of sacred groves. It may be presumed that it may have been included in the category of 'Village Forests,' (Regulation, 1891, Chapter 3, 28). If growth of agriculture, precipitated with population pressure, led to identification of sacred groves, then delineation began with the Regulation of 1891 wherein it stated that the State Government 'would specify the limits' of village forests. Even after the 46th year of India's Independence, in the Assam Forest Regulation (Amendment) Act of 1995, an amendment of the Regulation of 1891, the question of 'sacred groves' appeared to be a non-issue (Assam Gazette, 1995).

Another point to be noted is that with colonialism, along with the various forest regulation, came monetization and commercialization of agriculture that encouraged even the landlords and indigenous chiefs to extend cultivation, even if it meant clearing of sacred groves, and integrate themselves into the nexus of the Forest Department and the State Government. Introduction of Christianity was yet another factor that impinged on the local belief of the sacredness of forests, as in the case of Manipur mentioned earlier, in Meghalaya and Mizoram. Ritwick Dutta (2002) described the sacred groves of Meghalaya as a 'few pockets of undisturbed natural forests still in existence, comprising about 1000 sq. kms. protected by the 'tribals' and examines the factors responsible for the large-scale destruction of forests. Tiwari and others (1999) identified 105 sacred groves in Meghalaya. Under traditional forest classification of the Khasi-Jaintia, three types of forests *Law Lyngdoh*, *Law Kyntang* and *Law Niam* are declared as sacred groves in the Khasi, *Khloo Blai* in Jaintia hills, and *Asheng Khosi* in Garo hills. Some of these are owned by individuals, clans or communities, and are under direct control of the *Lyngdohs,* clan councils or local village *Dorbars/ Syiemships/ Dolloiships/ Nokmaships* (TNT 2017). These groves are dedicated to deities like *Ryngkew, Basa, labasa*. The groves abound in bamboo, needle wood, Indian birch, white pear, royal robe, balsum of peru, phurse champ, lac tree and plot's elm and several perennial streams originate from 58 of these groves. Ancestral worship in the shape of ancient monoliths is performed in the sacred groves (Tiwari et al. 1999).

These groves are protected under United Khasi-Jaintia Hills Autonomous District (Management and Control of Forests) Act 1958 and The Garo Hills Autonomous District (Management and Control of Forests) Act 1961, passed in pursuance of paragraph 11 of the Sixth Schedule of the Indian Constitution (Tiwari et al. 1995, 253-54). Dutta points out that a major factor that explains the diminish of sacred groves is the loss of 'sanctity' or 'sacredness' following urbanization, modernization, growth in population, introduction to Christianity, and the fact that no penal sanction in terms of imprisonment or fines for violation of rules have been laid down in the Forest Act of 1958 or also in Clause (b) Section 4 of the Management and Control of Forest Rules 1960, which deals with the management of sacred groves. Besides, Dutta states, 'the sacred groves are far too scattered to be regarded as one viable unit from the conservation point of view' and points out that even if brought under Protected Area management to preserve its biodiversity, sacred groves would lose their identity as being 'sacred' (Dutta 2002); Meghalaya's sacred groves bear the epithet of 'magnet for tourists.'

In pre-colonial Mizoram, every village had its own sacred grove known as *Ngawpui*, a natural habitat of streams, rocks, hills, plants and trees. These varied in size from a few to hundred hectares, from village to village. Land belonged to the community under the headship of the chief assisted by a group of elders, *Lal Upa*, and natural boundaries such as rivers and hills demarcated each village. Hence, each village was self-sufficient in terms of resources for the natural vegetation preserved for many generations provided food and natural products to the community. The Mizos also believe that deities like *Ramhuai, Lasi* and *Chawngtinleri* inhabit the sacred groves. For the Mizos, the sacredness of the plants is the reflection of the unity of life in nature and generates a sense of communion with the divine as the centre and source of life. Thus, the sacredness of the trees is an important facet of the religious, cultural and economic life. Several taboos such as the non-felling of trees, cutting of twigs and leaves, removal of dead wood, bearing of evil thoughts and injury to animals sheltering in the groves are prevalent (Amirthalingam 2016). Thus, religion played a pivotal role in conserving the flora and fauna of the Mizo hills as opposed to the classification made by Darlong and Barik (1998). This enabled the Mizos to preserve the sacred groves with its genetic resources and species as 'invaluable repository of genetic diversity.' During British rule, much of the sacred groves came under the British Forest policy posing a serious threat to their very existence while the introduction of Christianity and the decline of traditional beliefs and practices led to clearing of the sacred groves for construction of churches. However, the Mizos retained the sanctity of a few sacred groves of the pre-Christian period, such as the bamboo reserves called *Mawmund* in the Sialkal region of northeastern Mizoram (Malhotra 1990; Gokhale et al., 1998). In the post-Independence period, the Mizoram Forest Reserve Act of 1955 ensured that the sacred groves provide the basic livelihood needs of the village community and help in the upkeep of biological diversity for the future generations. The Mizos have *safety reserves* and *supply reserves* around the villages. These safety forests are a continuation of the sacred groves of the pre-Christian period (Joshi and Gadgil 1991; Amirthalingam 2016).

Sacred groves, also called *Gompa* Forest Areas (GFAs), of Arunachal Pradesh and Sikkim are mostly under the management of Buddhist monasteries. The ethnic communities of West Kameng, Tawang, Siang and Lower Subansiri districts of Arunachal have conserved the flora and fauna out of respect and belief in the power of nature and some are dedicated to local deities such as *Ubro* or *Ubram* and *Thouw-gew*. While Buddha was the deity in most of the sacred groves, many were anonymous and one had a Hindu deity in the form of Parbati/ Sati. A few of the sacred groves managed by Lamas and the Monpas, are attached to the Buddhist monasteries. A total of 65 sacred groves have been documented so far. These monasteries with sacred groves are mainly located in West Kameng and Tawang districts of the state; 58 GFAs were reported from these two districts and a few sacred groves from Lower Subansiri and Siang district of the state. Some of the species commonly found in the sacred groves of Arunachal Pradesh are banyan, pipal, *ashoka, bela* and *harada* (Khan et al. 2007). In Sikkim, 56 sacred groves have been identified and dedicated to deities like *Cho Chuba*, *Loki Sharia*, *Guru Padmasambhava* and *Rolu Devi*. The commonly found plant species are cupressus, silver oak, *tooni, thotnay, aiselu, tusare, sanu khari* and *ruk saro* (Amirthalingam 2016; Government of Sikkim 2011). Demojong pandam below Kanchenjunga in Sikkim is considered so sacred, that the people strongly believe that any human activity would spell disaster for the state (Khan et al. 2007).

Not much is known of sacred groves in Nagaland. The only mention is made by Manjil Hazarika but he does not mention any specific grove but in general describes them as the abode of 'sylvan spirits' who are appeased with the offering animals and eggs as sacrifice and several taboos such as felling of trees, chopping off branches, twigs and leaves or even the removal of dried wood, as with most sacred groves, are associated with it (Hazarika 2015, 36-37). Arkotong Longkumer (2022) who has done extensive work on the forests of Mokokchung states that sacred groves are not common in the district but he sees an overlap of the sacred and the profane. This is perhaps true for most of Nagaland. Similarly, as far as Tripura is concerned, the only reference has been made by Dipak Das (2020) who though describes Unakoti Hill as a sacred grove, emphasizes on its significance for butterfly diversity and as a biodiversity hotspot.

Sacred groves of Assam are mostly managed by ethnic communities like the Bodo and Rabha who term them as *Than* while it is *Madaico* for the Dimasa; *Madaico* is generally smaller in size with an area less than one acre. The number of groves is debatable as some research show as 29, of which 17 have been reported in Karbi Anglong district and 12 groves from North Cachar district (Medhi and Borthakur 2013) while Amirthalingam (2016) mentions 40 Sacred groves of Vaishnav monasteries or Sankaradeo maths as well, in the State which are home to the giant bamboo, pear bamboo, pink banana, metico pepper, Indian smilax, areca nut, sand paper tree and devil's cotton , which are considered as sacred plants. Sibrai, Alu Raja, Naikhu Raja, Wa Raja, Ganiyang- Braiyung and Hamiadao are the various deities to whom these sacred groves are dedicated. These groves are associated with the identity of communities and plant resources. The killing of animals and birds during the mating season is prohibited in and around the groves (Medhi and Borthakur 2013; Upadhyay 2019, 592). One of the largest sacred groves of Assam is Ronghang Rongbong, Hamren of Karbi Anglong District covering 10,434 sq. kms, which accounts for 13.3% of the total geographical area of the state. It mostly consists of, about 85 percent of the district, undulating and hilly terrain with numerous rivers and streams, while plain areas consist of valleys of the Jamuna, Kapili and Dhansiri rivers lying to the east. Thus, the landscape itself suggests greater bio-diversity in terms of trees, endangered species, medicinal plants etc. (Government of Assam 2022). Medhi and Borthakur (2013, 69) mention *Aludikho, Longmailaidikho, Longmailumdikho, Manjadikho, Damadidikho/Riaodikho, Hamridikho, Misimdikho, Baiglaidikho/Baigiadikho, Waibradaikho, Mongrangdikho/ Semkhordikho, Mongrangdikho, Ronchandidikho (Bamin)* as the 12 sacred groves of the Dimasas from North Cachar Hills District of Assam

Conclusion

Thus, the culturalization of nature in the form of sacred groves in Northeast India, reveals complex relationships between humans and nature or the physical world per se, society and cultural values, personal ties and other determinants such as political intervention, with socially constructed boundaries which are fluid and dynamic and therefore subject to historical change. This understanding hinges on the question: How 'sacred' are the sacred groves? In most cases, the 'sacredness', as the abode of the supernatural, deities and ancestor spirits, was a historical phase, though rituals and festivals still continue in some of the groves. Apart from resource use and viewed through multiple ethnic lens, sacred groves not only provide a differentiated view of local knowledge system but bears a tangible cultural heritage as well. The concept of societal relationships with nature or culturalization of nature, asserts that sacred groves are a consequence of historical as well as political processes, and hence are secular in more ways than one.

Urbanization with its accompanying so called 'development' and social transformation have had a large impact on the sacred groves and are responsible for their destruction. Construction of buildings, railways, roads, hydro-electric projects etc. have destroyed many groves in the past, while resources have become limited and activities like grazing, encroachment, cutting and collection of fuel wood, fruits and leaves have been increasing with social transformation, posing a threat to the conservation of biodiversity and ecosystem of the groves (Upadhyay 2019). This understanding explains the transition in the significance of the sacred groves of Northeast India, from 'Power' of nature to something sacred, to a site of economic exploitation under British rule, to establishing linkage as cultural heritage to ethnic movements in the late twentieth century, to being currently identified as hotspots for conservation of biodiversity and ecology and promotion of tourism. This transition itself questions the 'sacredness' of sacred groves. It underlines Gadgil's hypothesis that spiritual beliefs were no longer sufficient to ensure their survival *vis a vis* development and urbanization. Entrenched in the natural world, the preservation of these sites becomes central and meaningful to the indigenous cultures as cultural heritage that is strongly linked to the notion of territoriality, 'as modes of access to the past', and as Susan Davies (2008) maintains, making the connection between history and heritage, fundamental and inextricable'. Salute the 'Power' of the sacred groves in all their intensity and versatility!

References

Amirthalingam, P.M. 2016. “Sacred Groves of India – An Overview.” *International Journal of Current Research in Biosciences and Plant Biology*. 3(4) April: 64-74. Accessed October 11, 2022. http://dx.doi.org/10.20546/ijcrbp.2016.304.011.

Berghoefer, U., R. Rozzi, and K. Jax. 2010. “Many Eyes on Nature: Diverse Perspectives in the Cape Horn Biosphere Reserve and their Relevance for Conservation.” *Ecology and Society* 15(1): 18. Accessed May 11, 2021.
URL: http://www.ecologyandsociety.org/vol15/iss1/art18/.

Bhuyan, Ratna. 2019. “A Review Note on Shifting Cultivation in Northeast India amidst Changing Perceptions.” *Dhaulagiri Journal of Sociology and Anthropology* 13: 90-95. Accessed February 12, 2020. https://orcid.org/0000-0002-0494-5429.

Brandis, Dietrich. 1921. [4th Impression]. *An Account of Trees, Shrubs, Woody Climbers, Bamboos and Palms: Indigenous or Cultivated in the British Indian Empire.* London: Constable and Company Ltd.

C.P.R. Environmental Education Centre. Chennai: Ministry of Environment, Forests & Climate Change. Government of India. Accessed October 11, 2022. http://www.ncbi.org.in/.

Chandrakanth, M.G. and Jeff Romm. 1991. “Sacred Forests, Secular Forest Policies and People's Actions 31.” *National Resources Journal* 741. Accessed November 11, 2021.
https://digitalrepository.unm.edu/nrj/vol31/iss4/2.

Chouin, Gérard. 2002. “Sacred Groves in History: Pathways to the Social Shaping of Forest Landscapes in Coastal Ghana." *IDSIDS Bulletin* 33(1).

Dadosky, John D. 2004. *The Structure of Religious Knowing: Encountering the Sacred in Eliade and Lonergan*. Albany: State University of New York Press. Accessed February 23, 2020.
https://sunypress.edu › download › version › file › 9.

Darlong, V.T. and S.K. Barik. 1998. “Traditional Practices and Recent Initiatives in Biodiversity Conservation in Northeast India; Lessons from Peoples Experiences.” Seminar on: *Environmental Problems in North Eastern India.* Assam University, Silchar, India: Department of Ecology.

Das, Dipak. 2020. “The Butterfly *(Insecta: Lepidoptera*) Diversity of Unakoti Hill: An Important Sacred Grove of Tripura, North East India.” *Journal of Wildlife Research* 8(4) October- December: 81-85.

Davies, Susan. 2008. "History and Heritage." In: Peter Lambert and Phillip Schofield. Eds. *Making History: An Introduction to the History and Practices of a Discipline*. New York: Routledge. p. 280.

De Riencourt, Amaury. 1989. *Woman and Power in History,* New Delhi: Sterling Publishers.

Dutta, Ritwick. 2002. *Community Managed Forests: Law, Problems and Alternatives.* Accessed June 11, 2021. https://dlc.dlib.indiana.edu/dlc/bitstream/handle/10535/446/duttar020402.pdf?sequence=1.

Ecological Systems - Management Practices and Social Mechanisms for Forest and Climate Change in Uttarakhand. Accessed November 11, 2022. https://fsi.nic.in/forest-report-2021.

Gadgil, Madhav and V. D. Vartak. 1975. "Sacred Groves of India- A Plea for Continued Conservation." *Journal of the Bombay Natural History Society* 72(2): 314-320.

Gadgil, Madhav. 2018. "Sacred Groves: An Ancient Tradition of Nature Conservation." *Scientific American* 319(6) December: 48-57. Accessed August 10, 2019. doi:10.1038/scientificamerican1218-48. https://www.scientificamerican.com/article/sacred-groves-an-ancient-tradition-of-nature-conservation.

Glaser, Marion. 2006. "The Social Dimension in Ecosystem Management: Strengths and Weaknesses of Human-Nature Mind Maps." *Human Ecology Review* 13(2): 122-142.

Hazarika, Manjil. 2015. "Sacred Groves: Linking Conservation with Religion in North East India." *EcoWhisper* 2(1) March: 33-38. Garhwal, Srinagar: Society for Conserving Planet and Life (COPAL).

Hughes, D.J. and S.M.D. Chandran. 1997. Paper Presented in the Workshop on: *Role of Sacred Groves in Conservation and Management of Biological Resources*. Peechi, India: KFRI, cited in Sushma Singh, Mudasir Youssouf, Zubair A. Malik, and Rainer W. Bussmann. "Sacred groves: Myths, Beliefs, and Biodiversity Conservation—A Case Study from Western Himalaya, India." *Hindawi International Journal of Ecology* 2017(9): 1-12. Accessed October 12, 2021. https://doi.org/10.1155/2017/3828609.

Jaiswal, Vidhan. 2010. "Culture and Ethnobotany of Jaintia Tribal Community of Meghalaya, Northeast India - a Mini Review." *Indian Journal of Traditional Knowledge* 9(1): 38-44.

James, George Alfred. 1995. *Interpreting Religion: The Phenomenological Approaches of Pierre Daniel Chantepie de la Saussaye, W. Brede Kristensen and Gerardus van der Leeuw.* Washington: Catholic University of America Press.

Jeeva, S., B.P. Mishra, N. Venugopal, L. Kharlukhi, and R.C. Laloo. 2006. "Traditional Knowledge and Biodiversity Conservation in the Sacred Groves of Meghalaya." *Indian Journal of Traditional Knowledge* 5(4) October: 563-568.

Khan, M.L., A. Arunachalam, and A.R. Barbhuiya. 2007. *Web-GIS Digital Atlas of the Sacred Groves of the North-East India: Pilot Study with Sacred Groves of Arunachal Pradesh*. Government of India: Department of Forestry North Eastern Regional Institute of Science and Technology. [For Department of Scientific and Industrial Research Ministry of Science and Technology].

Khan, M.L., Ashalata Devi Khumbongmayum, and R. S. Tripathi. 2008. "The Sacred Groves and their Significance in Conserving Biodiversity: An Overview." *International Journal of Ecology and Environmental Sciences* 34(3): 277-291. New Delhi: National Institute of Ecology.

Khumbongmayum, A.D., M.L. Khan, and R.S. Tripathi. 2004. "Sacred Groves of Manipur, Ideal Centres for Biodiversity Conservation." *Current Science.* 87(4): 430-433.

Khumbongmayum, A.D., M.I. Khan, and R.S. Tripathi. 2005. "Sacred Groves of Manipur, Northeast India: Biodiversity Value, Status and Strategies for their Conservation." *Biodiversity and Conservation* 14: 1541–82.

Krishna, Nanditha and M. Amirthalingam. Eds. 2014. *Sacred Groves of India: A Compendium*. Chennai: C.P.R. Environmental Education Centre.

Longkumer, Arkotong. 2022. Senior Lecturer in Modern Asia. University of Edinburgh, and Senior Research Fellow at the Kohima Institute, Nagaland, through [Email correspondence with author on November 30, 2022.]

Malhotra, Kailash C., Yogesh Gokhale, Ketaki Das. 2001. "Sacred Groves of India: An Annotated Bibliography." *Indian National Science Academy and Development Alliance*. New Delhi: National Committee for Scientific Committee on Problems of Environment.

Medhi, P. and S.K. Borthakur. 2014. "Sacred Groves and Sacred Plants of the Dimasas of North Cachar Hills of Northeast India." *African Journal of Plant Science* 7(2): 67-77. Accessed October 10, 2020. http://www.academicjournals.org/AJPS. DOI: 10.5897/AJPS12.038.

Northeast India - A Mini Review. 2010. *Indian Journal of Traditional Knowledge* 9 (1) January: 38-44.

Ormsby, Alison. 2013. "Analysis of Local Attitudes Toward the Sacred Groves of Meghalaya and Karnataka, India." *Conservation and Society* 11(2): 187-197.

Oviedo, Gonzalo, Sally Jean Renaud, and Mercedes Otegui. 2005. *Protecting Sacred Natural Sites of Indigenous and Traditional Peoples: an IUCN Perspective.* Gland. Switzerland, Accessed June 11, 2021. https://www.jstor.org/stable/24707584.

Sacred Groves, Ronghang Rongbong, Hamren. 2022. Government of Assam Karbi, Anglong District: Forestry Information on Forest Management of Karbi Anglong District. Accessed February 2, 2022. https://Karbianglong.Gov.In/Departments/Forestry.

Schultz, Irmgard. 2006. "The Natural World and the Nature of Gender." In: K. Davis, M. Evans, and J. Lorber. Eds. *Handbook of Gender and Women's Studies.* London: Sage Publications. pp. 376-96.

Schwartz, A. 2014. "Sacred Space." In: D.A. Leeming. Eds. *Encyclopaedia of Psychology and Religion,* 1579–1583. Boston, MA: Springer. Accessed May 14, 2022. https://doi.org/10.1007/978-1-4614-6086-2_9001; https://link.springer.com/referenceworkentry/10.1007/978-1-4614-6086-2_9001.

State of Forest Report: 2021. Dehradun, India: Forest Survey of India, Ministry of Environment and Climate Change.

Sukumaran, S., A. Pepsi, D.S. Siva Pradesh, and S. Jeeva. 2018. "Phytosociological Studies of the Sacred Grove of Kanyakumari District, Tamilnadu, India." *Tropical Plant Research* 5(1): 29–40.

The Assam Forest Regulation of 1891. Accessed October 22, 2021. https://forest.assam.gov.in/sites/default/files/swf_utility_folder/departments/pccf_lipl_in_oid_4/menu/document/assamForestRegulation1891.pdf.

The Assam Gazette Extraordinary. 1995. Guwahati: Government of Assam. pp. 556-563.

The Forest, Environment and Wildlife Management Department. 2011. Gangtok: Government of Sikkim.

The Northeast Today (TNT). June 5, 2017. Accessed November 20, 2022. https://thenortheasttoday.com/states/arunachalpradesh/5-sacred-groves-of-the-northeast/cid2532272.htm

Tiwari, B.K., S.K. Barik, and R.S. Tripathi, 1999. *Sacred Forests of Meghalaya: Biological and Cultural Diversity*, Shillong: Regional Centre, National Afforestation and Eco-Development Board, North Eastern Hill University, Shillong.

Upadhyay, Keshav Kumar, Bakerbha Japang, Ngangbam Somen Singh, and S.K. Tripathi. 2019. "Status and Socio-ecological Dimensions of Sacred Groves in Northeast India." *Journal of Applied and Natural Science* 11(3): 590-595. Accessed November 23, 2021. https://doi.org/10.31018/jans.v11i3.2121.

Notes on Contributors

Alison A. Ormsby teaches Environmental Studies at the University of North Carolina Asheville, USA. She is a human ecologist with 30 years of experience working on topics including people and protected areas, environmental education, and sacred natural sites. She has conducted research on sacred forests in Sierra Leone, Ghana, and India. She is a member of the IUCN's Specialist Group for Cultural and Spiritual Values of Protected Areas, and has numerous publications on her work.

Antaryami Pradhan is an Ecologist with eight years of experience working on biodiversity conservation and ecosystem services in Odisha. He has conducted research on sacred natural sites in Western Odisha, and engaged in urban ecological studies in Bhubaneswar, biodiversity conservation studies in protected areas of the Odisha, including the Bhitarkanika mangroves. He has published many research works on biodiversity conservation and ecological services in Odisha, India.

Jonathan C. Onyekwelu is a Professor of Silviculture and Forest Management at the Department of Forestry and Wood Technology, Federal University of Technology, Akure, Nigeria, where he has been a faculty member since 1992. He obtained his doctoral degree from Technical University of Munich, Germany. Professor Onyekwelu is a German Academic Exchange Programme (DAAD) and Alexander von Humboldt Foundation (AvH) fellow. His research interests are in the areas of silviculture, forest ecosystem restoration and management, biodiversity conservation, forest biomass and carbon storage, where he has published extensively. He has collaborated actively with researchers, to win, and successfully executed international research grants.

Laxmi Hansepi is pursuing Ph.D at the Department of History, Gauhati University, Assam, India. Her doctoral dissertation is titled, 'Festival, Culture and Identity of Karbis in Karbi Anglong District, Assam.' Her research interests are in Ethnic Identity, Festival Studies, Oral History and Community, and Women. Her recent publications in journals are on Karbi Women and Environmental Conservation (*Grassroots Journal of Natural*

Resources 2022), and Ritual of the Dead: Chomangkan Festival in Cultural Context *(Religions 2022).*

Madhusmita Samal has a Masters in Life Sciences from AIPH University, Bhubaneswar and is now working towards ecosystem services provided by urban biodiversity in major cities of Odisha. She has interest in researching and supporting the cultural benefits of the green spaces in and around major cities of Odisha.

Meeta Deka is Professor (Retd.) Department of History, Gauhati University, Assam, India. She is author of *Student Movements of Assam* (Vikas 1996), *Women's Agency and Social Change: Assam and Beyond* (Sage 2013) and *Urbanisms in South Asia: Northeast India Outside In* (Primus 2020).

Michael J. Sheridan teaches Anthropology, Environmental Studies, and African Studies at Middlebury College in Vermont, USA. His most recent publication is *Roots of Power* (Routledge 2023*)*, a multi-sited political ecological study of ethnobotanical institutions and social change in Tanzania, Cameroon, Papua New Guinea, French Polynesia, and St. Vincent.

Parismita Borah is a Research Scholar at the Department of History, Gauhati University, Assam, with specialisation in Ancient Indian History. Her research interests include religion, culture and environment, and gender dynamics in Northeast India. Her doctoral dissertation focusses on Traditional Knowledge Systems and Biodiversity Conservation. Currently she is working as an Assistant Lecturer at Aakash Educational Service Limited, Guwahati.

Ripun Doley is an Assistant Professor at the Department of History, Dakshin Kamrup Girls' College, Assam, India. His doctoral dissertation was titled, 'Religious Institution and Acculturation among the Misings of Majuli, Assam.' He specializes in Ancient Indian History. He has published many scholarly articles on Religion, Cultural Identity and related fields.

Rosy Yumnam is an Associate Professor at the Department of English Language Teaching, The English and Foreign Languages University, Regional Campus, Shillong. She is author of *Culture in English Language Teaching: The Use of Puyas in English Literature* (Mittal Publications 2021). Her research and publications address language, culture and pedagogy, teaching language skills, production of ELT materials and academic writing.

Sakti Sekhar Dash is Fellow of Social Science Research Council, Open Association of Research Society, USA. Currently, he is engaged as a Senior Researcher in the Department of English, Ravenshaw University, India. With a profound interest in literature, history, and culture, he has studied the myths, legends, and folklore from various parts of the world. An experienced educator, he has extensively written and published on a diverse range of topics, including, modernism, Greek drama, environmental studies, Theatre of the Absurd, and Shakespearean drama.